IMAGES
of America

OAK PARK AND THE MONTGOMERY ZOO

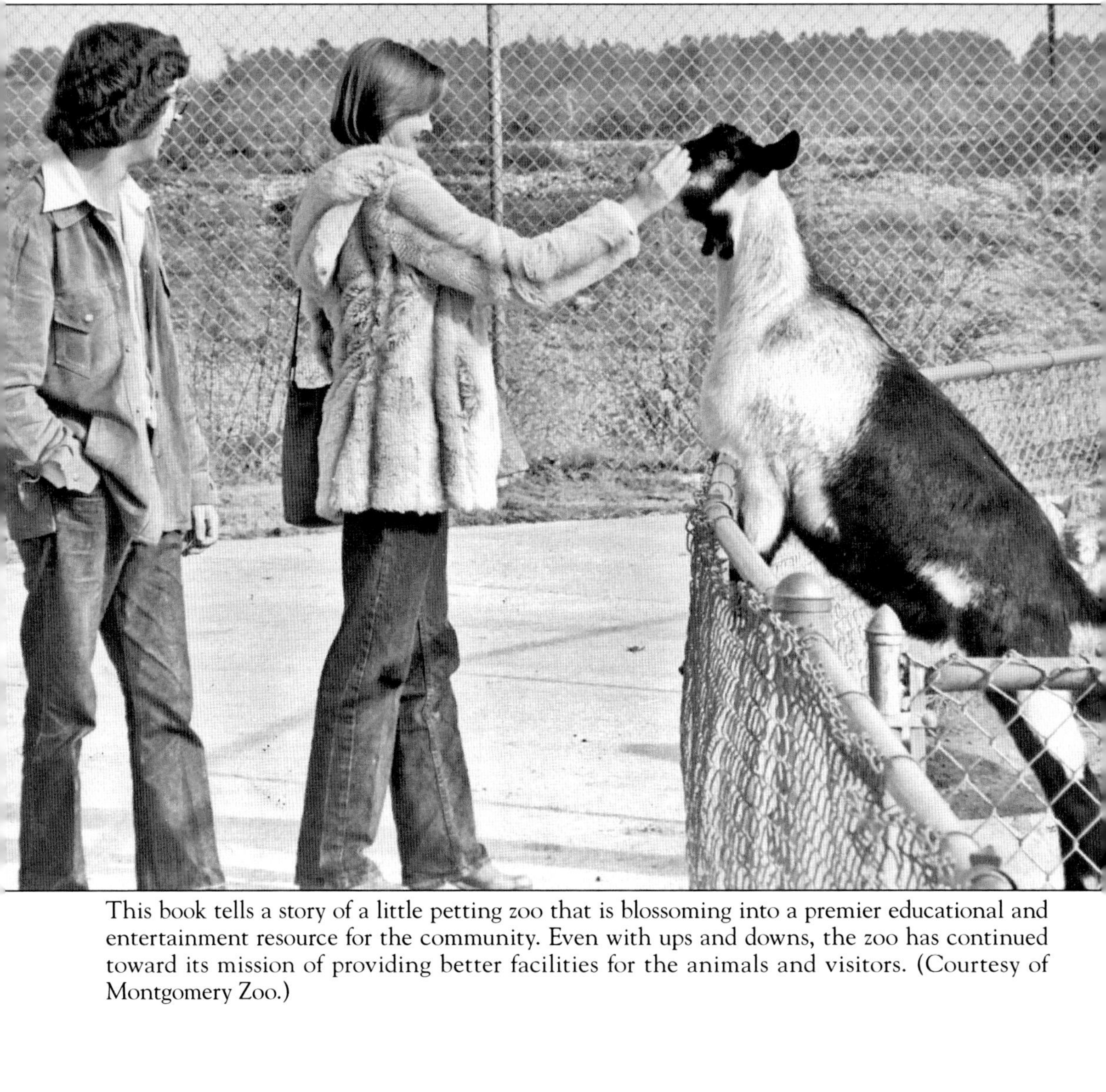

This book tells a story of a little petting zoo that is blossoming into a premier educational and entertainment resource for the community. Even with ups and downs, the zoo has continued toward its mission of providing better facilities for the animals and visitors. (Courtesy of Montgomery Zoo.)

ON THE COVER: With the construction of the main Oak Park pavilion with an encircling porch, the park became the major location for dances and parties. The large ornate wooden structure was nestled on a rise in the trees overlooking the winding paths, picnic areas, and playground. Unfortunately many of the early Oak Park records were tragically lost in the city hall fire of May 1932. (Courtesy of Robert Fouts/Fouts Commercial Photography.)

IMAGES
of America

OAK PARK AND THE MONTGOMERY ZOO

Heather S. Trevino and
Linda E. Pastorello

ARCADIA
PUBLISHING

ISBN 978-0-7385-5311-5

Published by Arcadia Publishing
Charleston SC, Chicago IL, Portsmouth NH, San Francisco CA

Printed in the United States of America

Library of Congress Catalog Card Number: 2007931380

For all general information contact Arcadia Publishing at:
Telephone 843-853-2070
Fax 843-853-0044
E-mail sales@arcadiapublishing.com
For customer service and orders:
Toll-Free 1-888-313-2665

Visit us on the Internet at www.arcadiapublishing.com

To Carole King of Old Alabama Town, Landmark Foundations. Without her valuable help, patience, and resources, we would not have been able to find and document the history of zoo. For only her interest as a historian, Carole dedicated countless hours toward researching and scanning photographs.

CONTENTS

ACKNOWLEDGMENTS

Special thanks to the Montgomery Zoo deputy director Marcia Woodard for providing us assistance and support to pursue such an important opportunity in documenting the history of the zoo; and to Carole King of Old Alabama Town, Landmark Foundations, for taking time out of her busy schedule to meet with us. Photographs were provided by Carole King, Mary Ann Neely, Robert Fouts Photography, Martha and Howard Green, Kitty Chamberlain, Florence Evans Jordan, Lula Mae Myers Chapman, Betty McDuffee, June Thompson, V. Norman Dennis, Fran and Sheila Bergquist, Betty Breslin, Richard Hammond, Jerry Smith, Marsha T. Lautenbach, Joe Migon, Roger Wentowski, Karen W. Carden, and Charlie Steed.

FOREWORD

The city of Montgomery is rich with history. Named for Maj. Lemuel Putnam Montgomery, the first man killed in the Battle of Horseshoe Bend in 1814, the city was selected as Alabama's state capital in 1846, a few years before the onset of the Civil War. The Alabama State Capitol Building in Montgomery also served as headquarters for the Confederacy and was the site of the inauguration of Confederate president Jefferson Davis. Montgomery boasts the first electric streetcar in the country and also the nation's first flight school. The Wright Flying School, from which Orville and Wilbur Wright conducted their first night flights, was established in 1910 on the site that is now Maxwell Air Force Base. In 1955, Montgomery made history when Rosa Parks was arrested for refusing to give up her bus seat to a white man and became a civil rights heroine. The reaction to this incident led to the 382-day Montgomery bus boycott, which finally forced the city to desegregate its transit system.

The cultural heritage of Montgomery is kept alive today, with many attractions throughout the city. In Old Alabama Town, one can discover over 100 years of history journeying through restored and furnished buildings representing the 19th and 20th century of central Alabama. The Montgomery Museum of Fine Arts, founded in 1930, is the oldest fine arts museum in the state. A popular tourist attraction is DeSoto Caverns Park, which houses one of the oldest recorded caves in the United States. Montgomery is also home to the Alabama Shakespeare Festival, which is the sixth largest Shakespeare festival in the world.

Also steeped with history is one of the most popular attractions of the city of Montgomery, the Montgomery Zoo, which can boast over 200,000 visitors a year. The Montgomery Zoo was established in 1935 as part of Oak Park. It grew and thrived there until 1959. Eleven years later, the zoo was reestablished and moved to its current located in north Montgomery. The Montgomery Zoo opened to the public at its current location in 1972 as a 6-acre facility. At that time, the animal collection consisted of 84 species acquired mostly as gifts from zoos and individuals.

A professional design team was organized in 1990 and given the task to design a master plan for a new zoo. The resulting 40-acre zoo was opened to the public in September 1991. Five continents with barrier-free, state-of-the-art exhibits provide visitors with exciting educational and interpretive opportunities throughout the zoo.

New exhibit construction and growth has been substantial since the opening in 1991. The most recently completed construction project was the opening of the African Elephant Habitat in 2005. Currently the River Otter Habitat is under construction in the North American Realm of the zoo. Since the opening of the new zoo in 1992, the Montgomery Zoo has grown by nearly 700 percent in acreage, more than doubled the size of the animal collection, created over 50 new jobs, and become a major attraction and economic asset in the community. The Montgomery Zoo recently made the state of Alabama's top-10 most visited attractions.

We strive to meet our mission statement of improving the future for wildlife by exhibiting animals and plants and providing education and conservation programs that encourage respect and stewardship of the natural world and a better understanding of our place within it. Our five major goals are:

1. The zoo will provide for the health and well-being of the animals in its care.
2. The zoo will be a leader and innovator in wildlife conservation.
3. The zoo will be an enjoyable year-round facility for all ages and income.
4. The zoo will have a high-quality and diverse conservation/education program.
5. The zoo's plant and animal collection will be available and utilized for study and research for better understanding of our natural environment.

The Montgomery Zoo has evolved to be an asset to the people living in the central Alabama region. We continue growing simply because the visitors that walk through our gates are committed to having a state-of-the-art zoological facility in their community. This book will provide many countless hours of history of the old zoo that we're sure will help you appreciate how we began, where we were, and the current status of the Montgomery Zoo. Royalties of the book will be donated on behalf of the authors to the Montgomery Zoo toward funding future projects.

Enjoy the read, and we hope to see you at the zoo.

—Doug Goode, Director
—Marcia Woodard, Deputy Zoo Director

One

Oak Park Zoo
1935–1959

Oak Park, established in the late 1800s with a total of 40 acres, was the social and recreational center of Alabama's capital city of Montgomery. It was here, in 1935, that a group of animals were housed in facilities built by the WPA and called Oak Park Zoo. The menagerie included monkeys, bears, alligators, deer, and a pair of lions, one purchased from a zoo in Arizona and the other a gift of the Ringling Brothers Circus to Mayor W. A. "Tacky" Gayle. Rounding out the zoo were two small train rides, a merry-go-round, a Ferris wheel, and a swimming/wading pool.

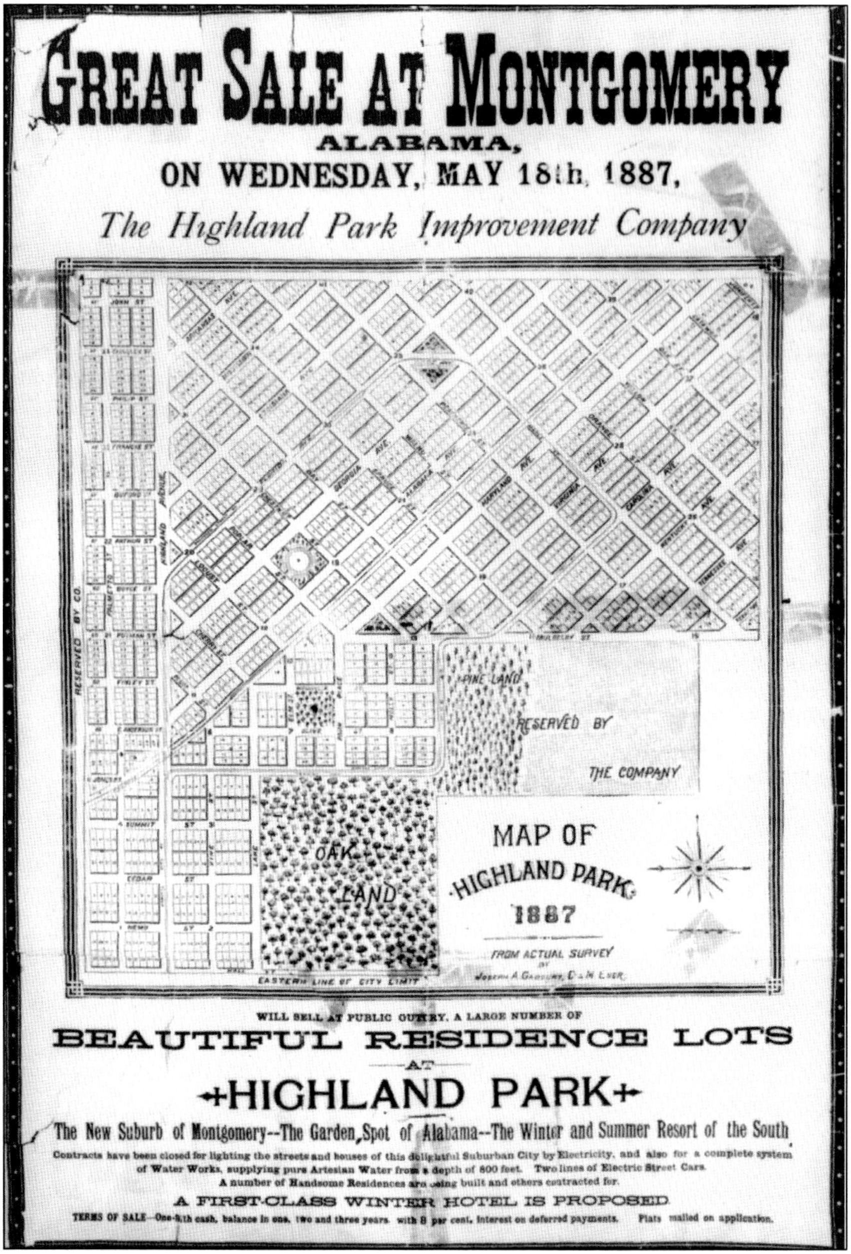

The directors of the Capital City Street Railway Company planned to add a rail line to Oakwood Cemetery. At that time, the cemetery was also used as the city park. The railway company purchased this property from Col. Bolling Hall, a planter and statesman, on December 21, 1886, for $40,000. The land, just a mile from Court Square, was bought to create a park. The Montgomery City Directory for 1887 reported the park as a "place where the public may spend leisure hours in pleasant drives and recreation [and] will have a splendid artesian lake and boats, a two and a half mile drive of tracks." Two electric streetcars provided visitors with transportation to and from what was then called Highland Park. The Montgomery City Council purchased 45 acres of the property from the Highland Park Improvement Company on May 9, 1899, for $25,000. (Courtesy of Robert Fouts/Fouts Commercial Photography.)

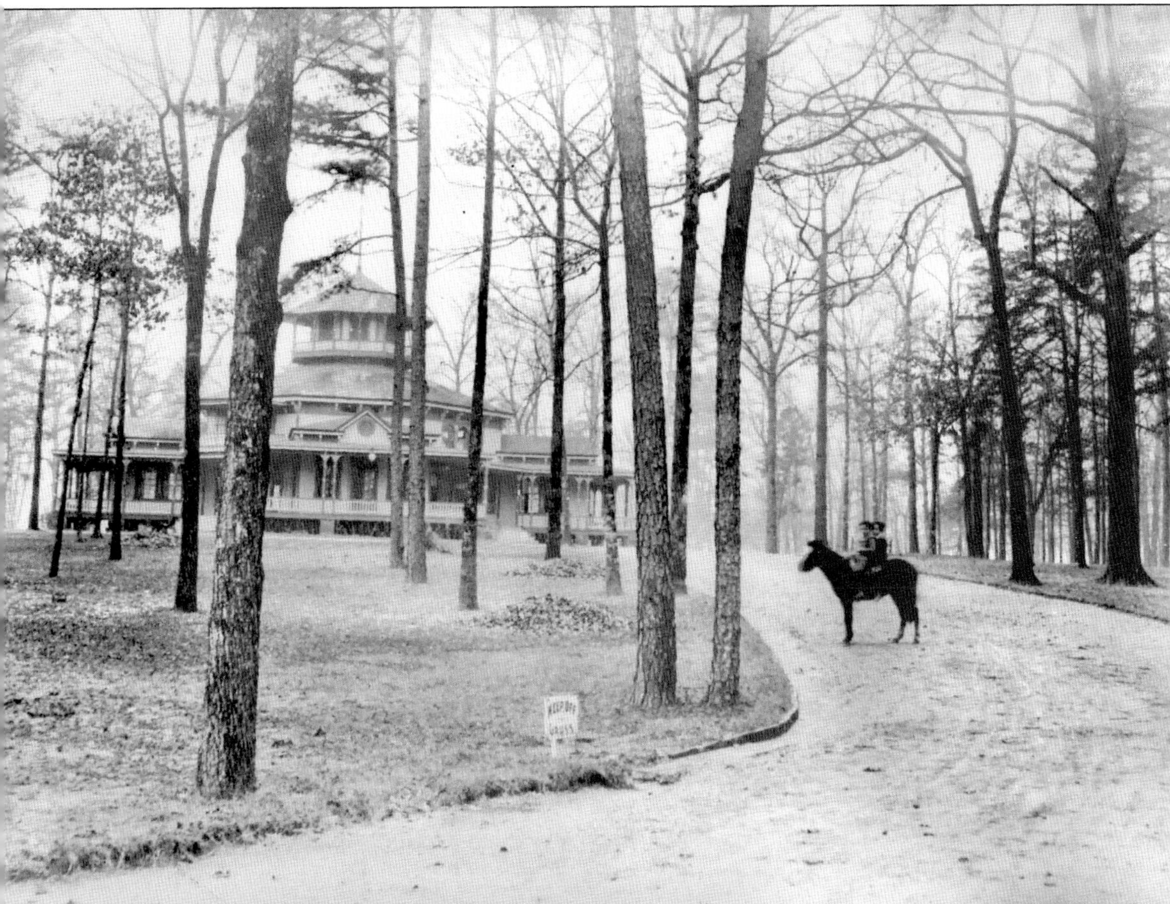

As the first overseer of the park, Joseph Armstrong Hudson collected $65 per month to survey the park from the back of his trusty donkey. Old Highland Park was at that time heavily wooded, with practically no flowers. The driveways through the area were bridlepaths rather than roads. Mayor John Hughes Clisby was a major supporter who persuaded the Montgomery city officials to purchase the land from the railway company. Because the councilmen and Montgomerians believed that the city spent too much money for the land, the park was known for years as "Clisby's Folly." In 1899, the Montgomery City Council changed the name of the new park from Old Highland Park to Oak Grove. (Courtesy of Robert Fouts/Fouts Commercial Photography.)

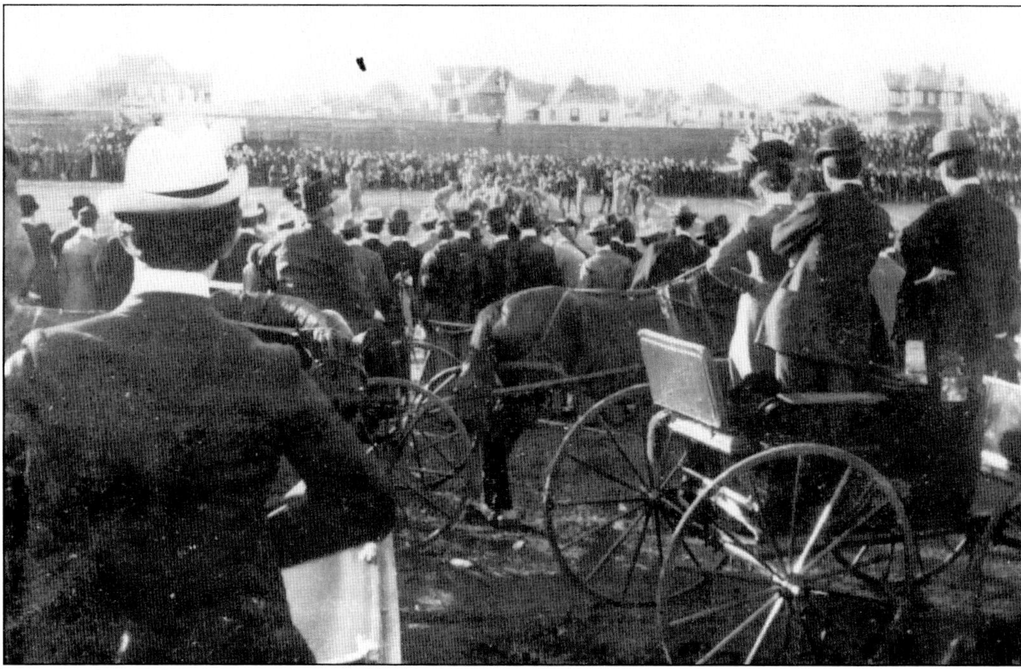

During the early 1900s, Oak Park was used for football games, where groups would gather by way of mule-drawn cars. Regional football teams used the park for their games including Auburn University, Cumberland University, Sewanee: University of the South, Clemson College, and University of Alabama. One of the most noted games was in 1896 between Auburn University and Sewanee with Auburn as the victor, 38 to 6. But in 1903, Sewanee shut out Auburn 47 to 0. Below is a 1904 varsity team photograph of one of the competing football schools. (Above, courtesy of Davidson/Landmarks Foundation; below, courtesy of Oglesby Collection.)

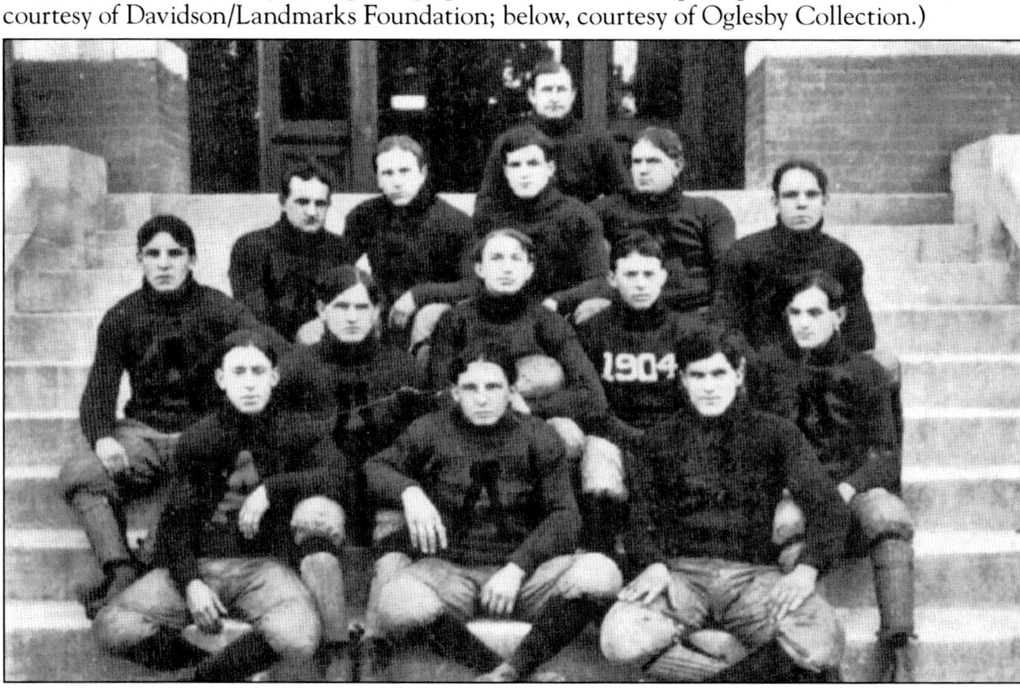

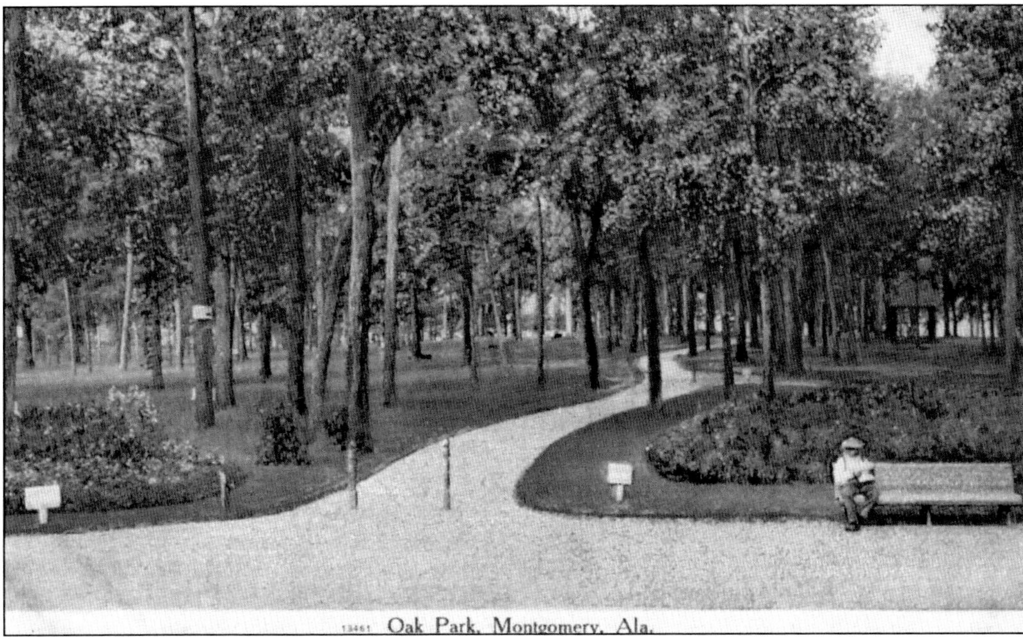

Oak Park was the center of social and recreational activities. During Easter, children and adults would come to the park dressed in their Sunday best. Local schools would bring their students to the park for square dancing, band concerts, and all-day singings. The pavilion was the center for boy-girl dances because it had its very own rockola, which would play songs for a nickel. (Courtesy of Kitty Chamberlain.)

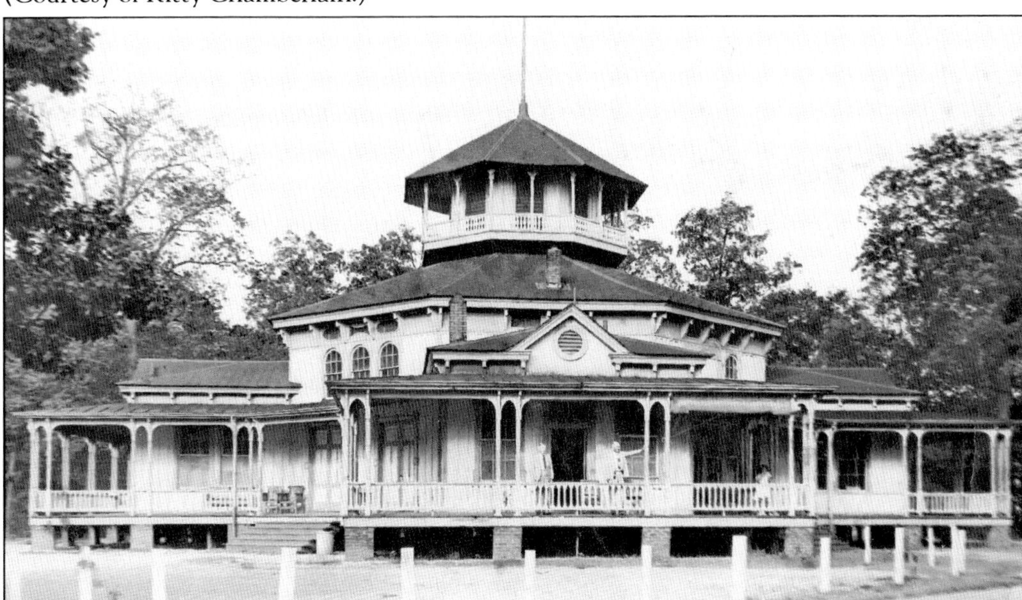

The construction of the main Oak Park pavilion, with an encircling porch, made the park the major location for dances and parties. The large ornate wooden structure was nestled on a rise in the trees overlooking the winding paths, picnic areas, and playground. Unfortunately many of the early Oak Park records were tragically lost in the city hall fire of May 1932. (Courtesy of Robert Fouts/Fouts Commercial Photography.)

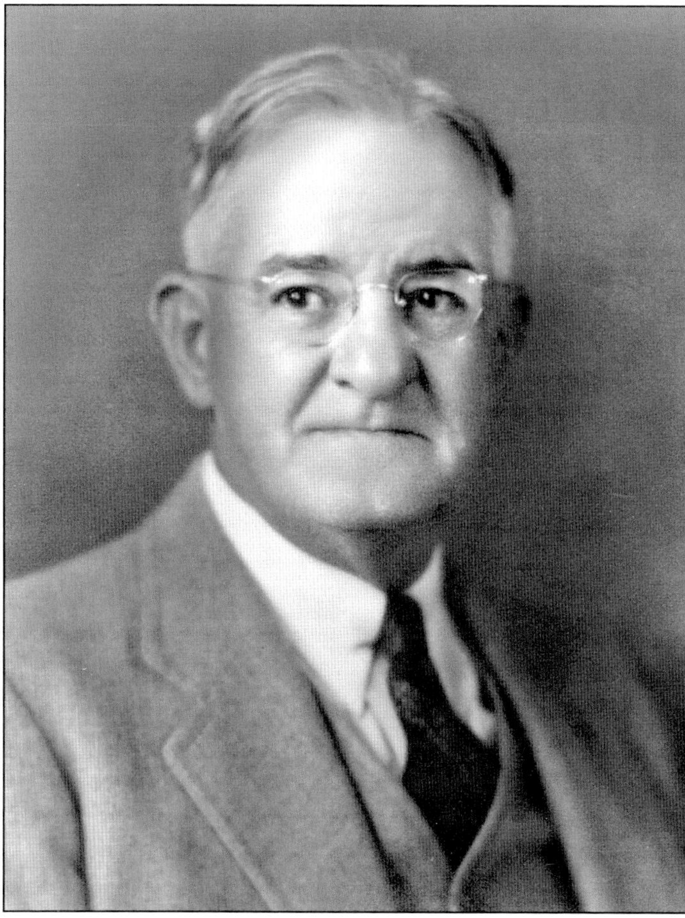

William A. Gunter Jr. was the mayor of Montgomery for 27 years between 1910 and his death in 1940. He is remembered for his many achievements in public service. Not so well known are the personal stories of lives he touched and the suffering he used his power to alleviate. During the Depression and before any public assistance was available, many people appealed to him for help. The humanitarian that he was, he found creative ways to put them on the city's payroll. Stories of the people he helped survive those hard times abound, but here is one from the late 1930s, where Oak Park was the staging ground. (Above, courtesy of Florence Evans Jordan (Mrs. George E.), granddaughter of Mayor William A. Gunter Jr.; below, courtesy of Kitty Chamberlain.)

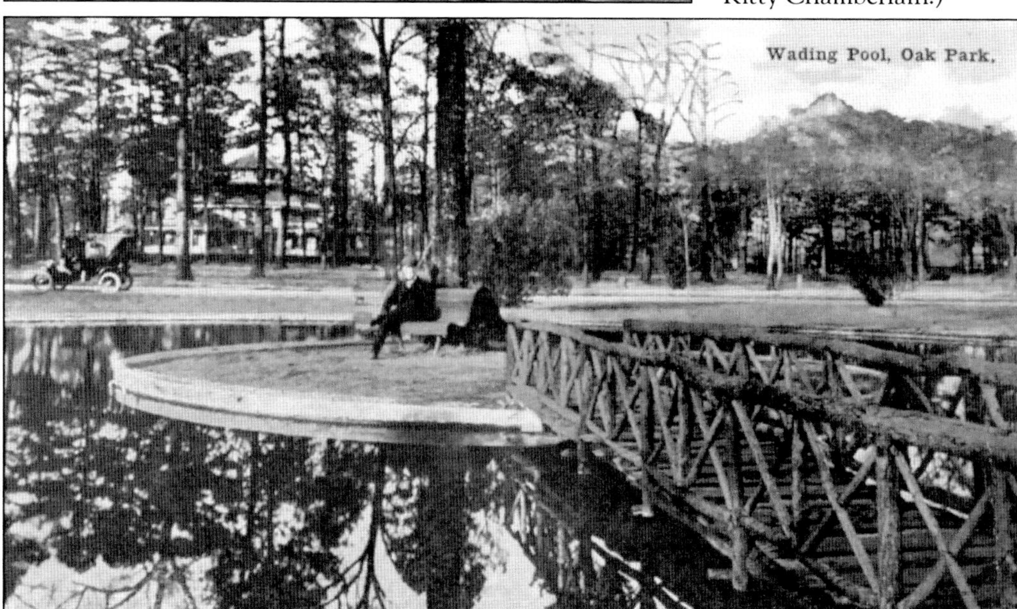

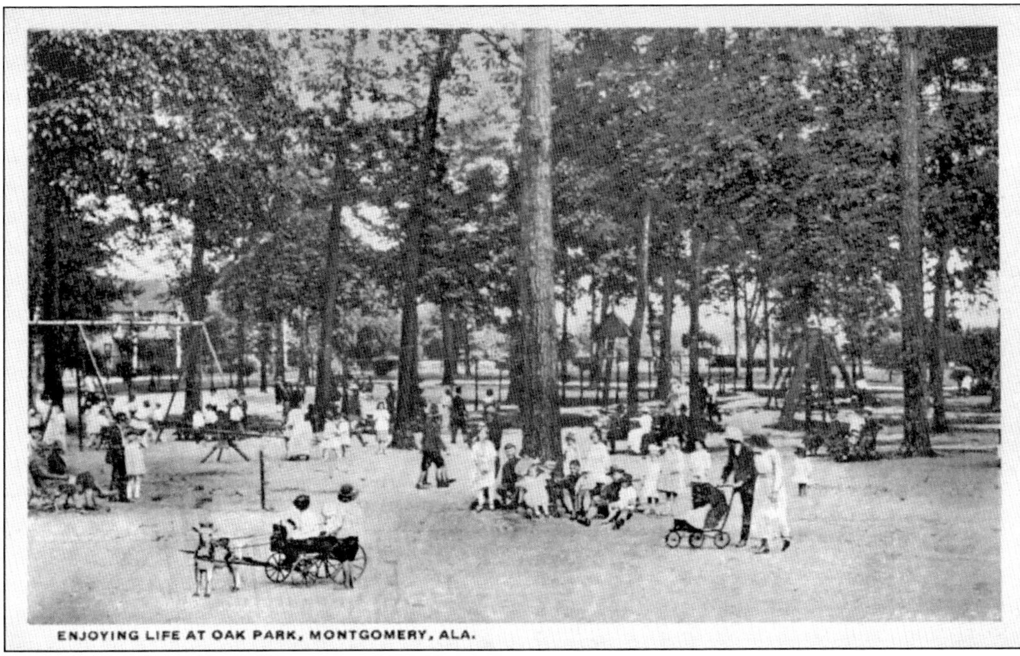

The park is bounded by Lake Street (north), Palmer Street (south), Hall Street (west), and Forest Avenue (east). Capital City Street Railway expanded eastward, allowing access to Oak Park. The trolley would travel the mile and a half from Court Square or from Highland Park Station. Trolley cars would travel through Oak Park and up Third Street to Mulberry Avenue, where they would reverse themselves and go back. (Courtesy of Landmarks Foundation.)

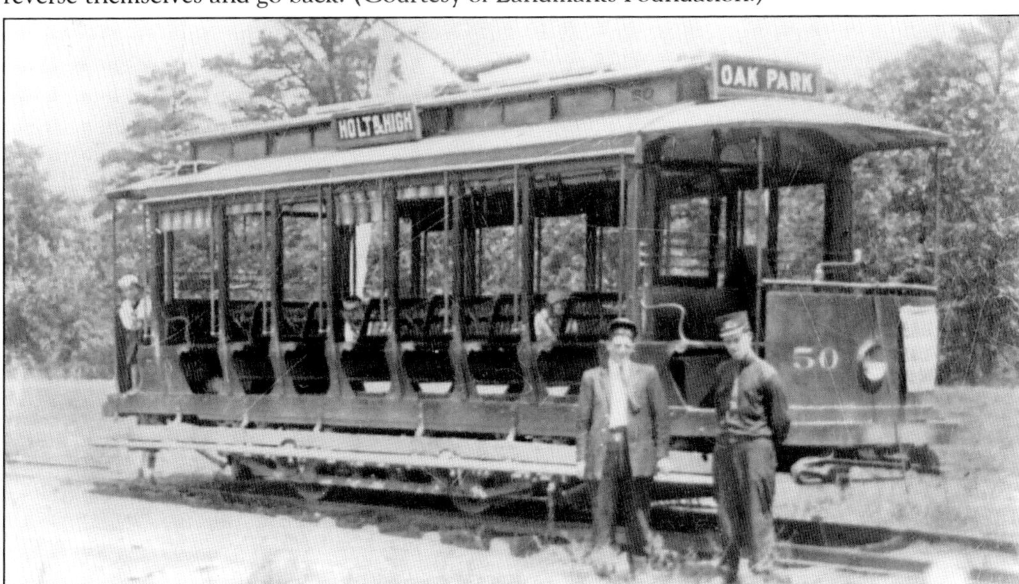

Claude DeYampert and her sister, Daisey DeYampert, who had fallen on hard times, were the last members of an old aristocratic family. Mayor Gunter provided the DeYampert sisters with jobs in the city parks. Daisey was taken to Oak Park every weekday during the summer, where she would sit under the shade of a large oak tree and tell stories to the local children. (Courtesy of Kitty Chamberlain.)

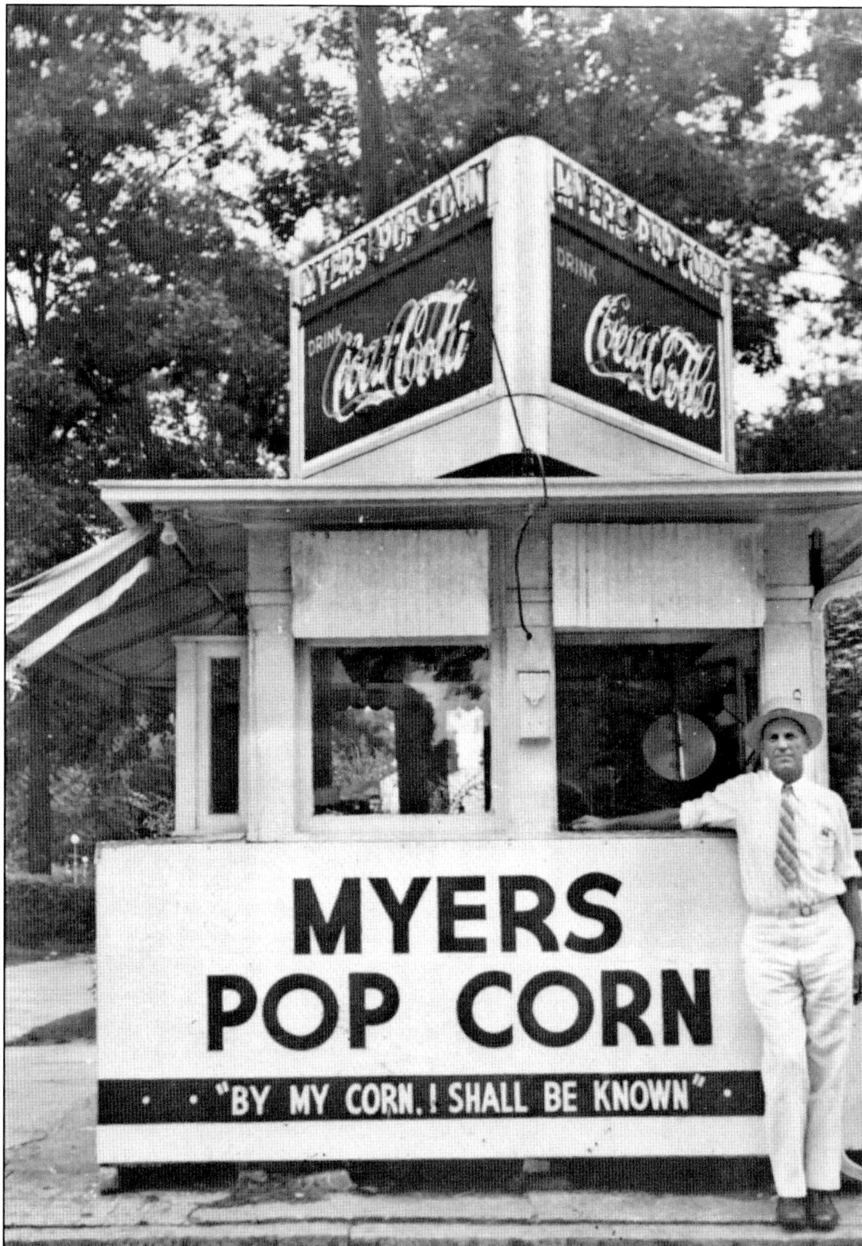

In July 1922, the Myers brothers opened two popcorn stands in different locations of Montgomery. A friend and neighbor, Mr. Bogacki, built M. B. (Pop) Myers a structure just outside the east gate of Oak Park on Forest Avenue from which to sell his popcorn. Pop and his family lived across the street in a large, white, frame two-story house, allowing his stand to become a permanent fixture in the area. Pop was best known for his white popcorn and his "shake-ice" cokes, an almost—but not quite—frozen Coca-Cola in a 6.5-ounce glass bottle. He was the only business to sell "white" popcorn instead of yellow corn. Pop's entire family worked the stand, with the older children manning the inside while the younger ones went from car to car taking orders. (Courtesy of Marcus Myers Pugh Foundation.)

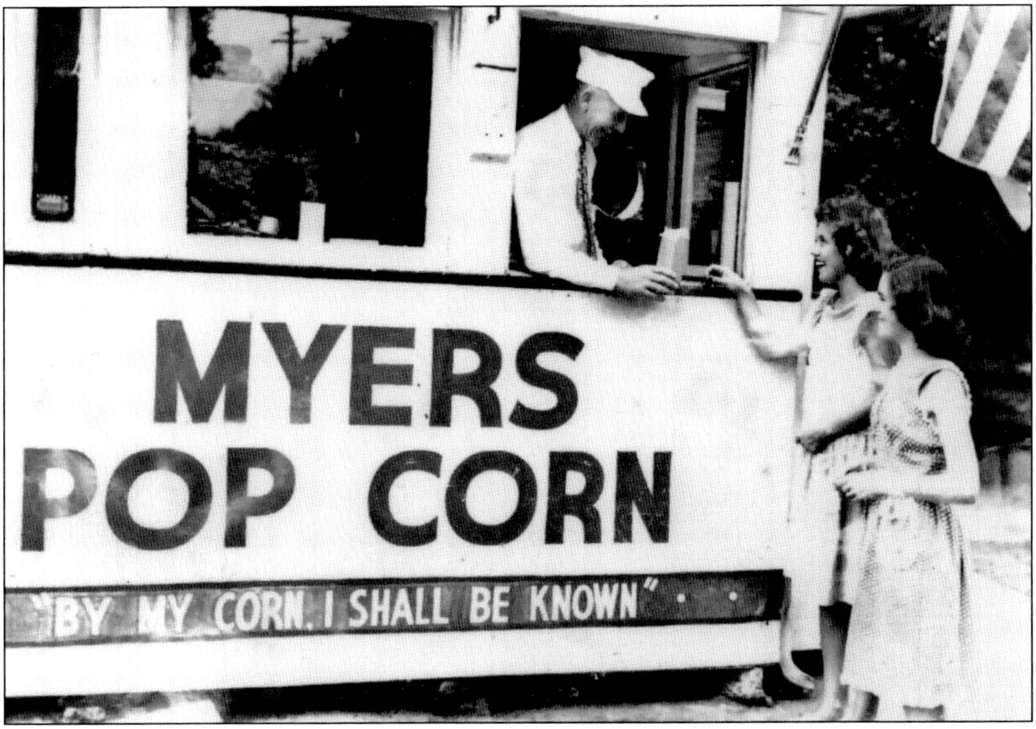

Pop Myers was also known for freezing chocolate bars, such as Milky Ways and Hershey's Bars. He kept a little notebook in which he would keep track of his younger customers and how much they purchased. On the weekends, their parents visited the stand to settle up their children's tabs. (Courtesy of Lula Mae Myers Chapman.)

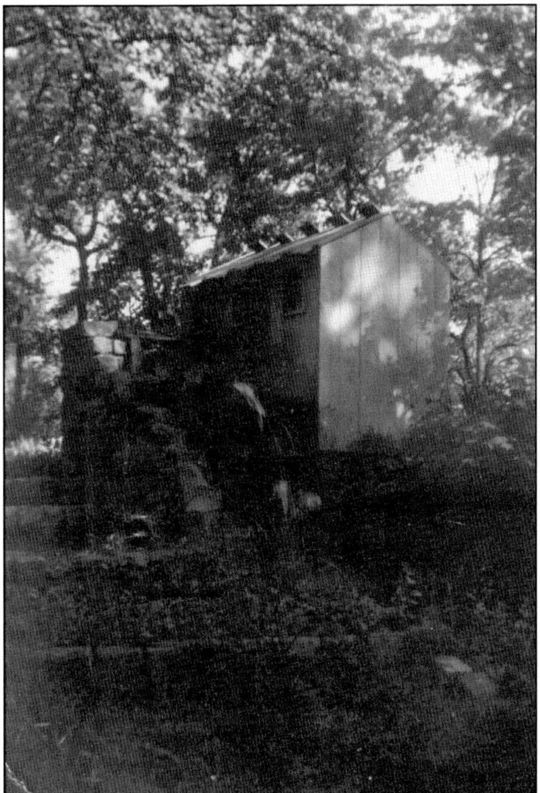

The water mill was a fixture in Oak Park, located near the animal cages and Monkey Island. Visitors would sit by the trickling stream to watch the animals or picnic in the shade. The water mill is still a visitor attraction today in Oak Park. (Courtesy of Cargill/McDuffee Collection.)

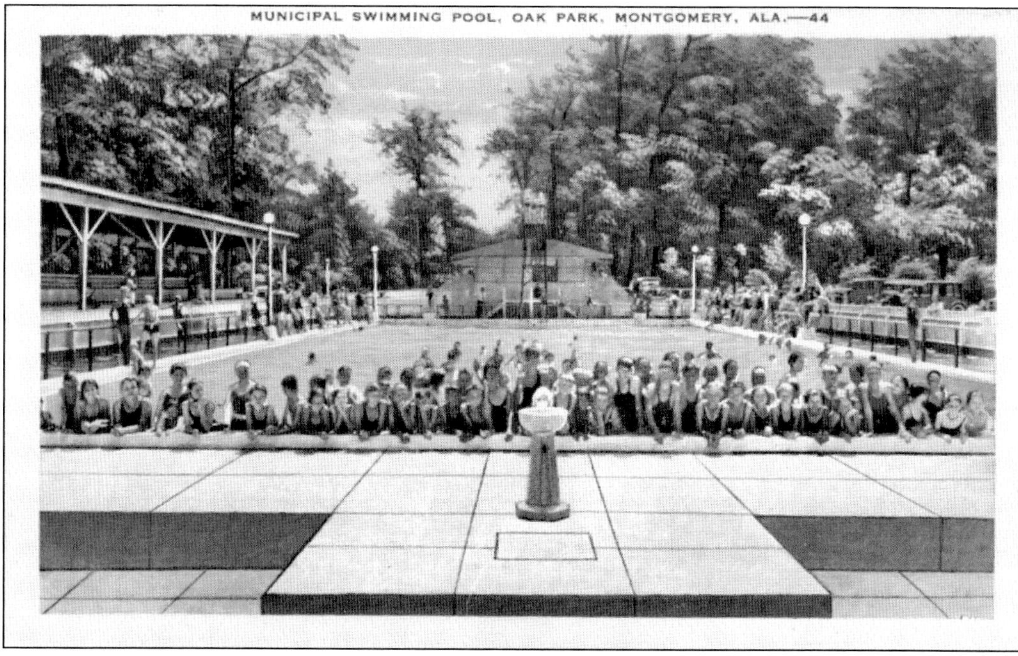

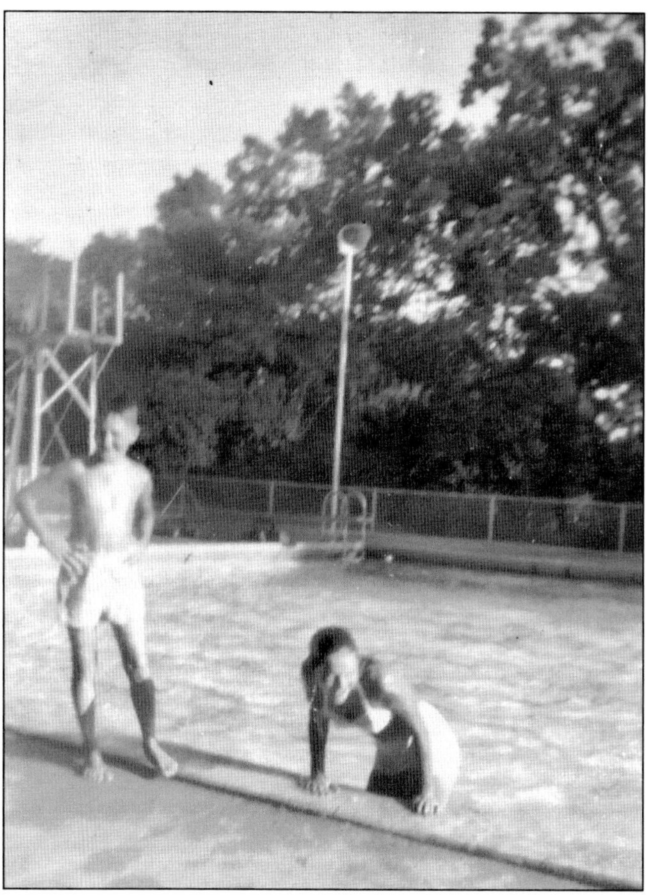

The swimming pool was the largest in the city, measuring 40 by 60 feet. One side of the swimming pool featured a covered grandstand, and a bathhouse was at the end of the pool. This new addition made Oak Park a very popular cooling-off spot for Montgomery's youth. (Courtesy of Kitty Chamberlain.)

Oak Park's pool was open from 6:00 a.m. until 10:00 p.m. Most of a child's time was spent swimming in the pool or in the wading pool. The a wading pool was for the smaller children and a larger swimming pool for the older children. However, the pool was closed several times between 1950 and 1954 because of possible polio outbreaks. (Courtesy of Cargill/McDuffee Collection.)

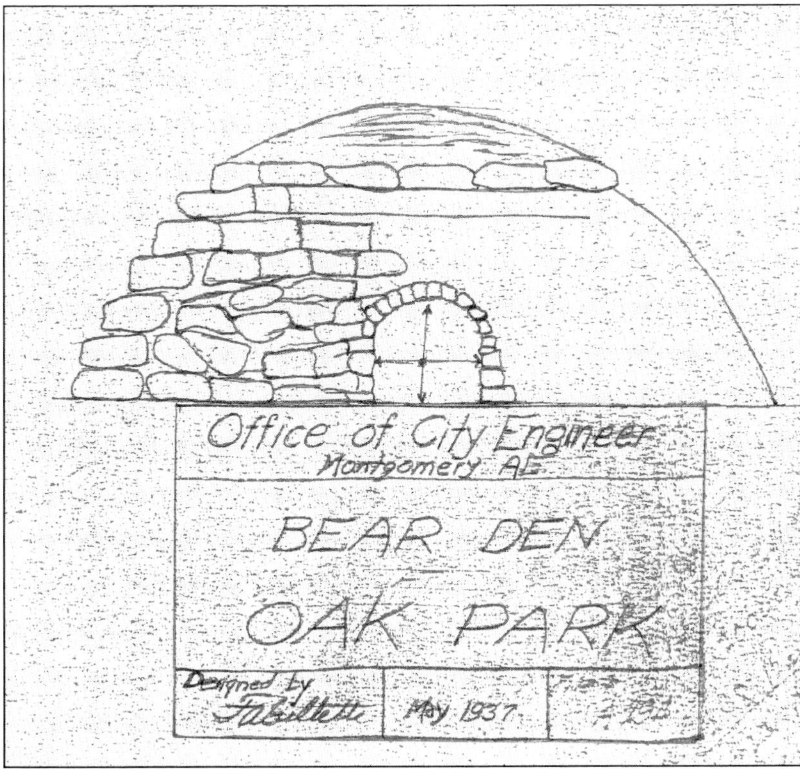

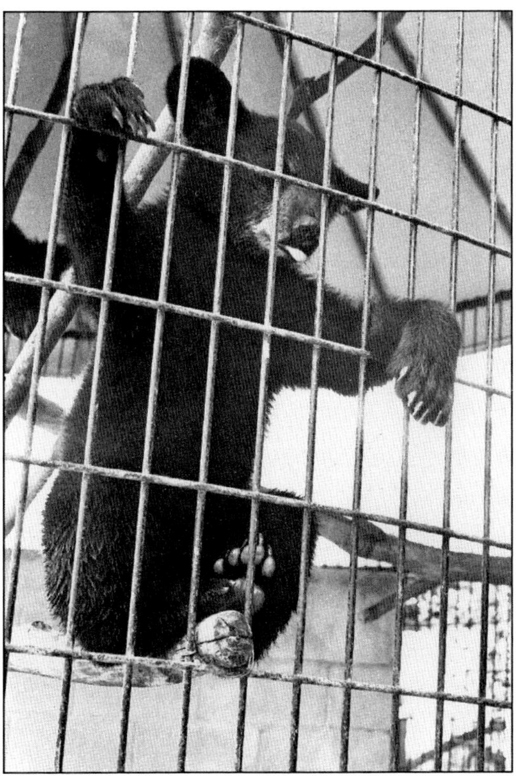

The landscape architectural firm of Olmstead Brothers produced a plan for improvements to Oak Park that included a swimming pool, tennis courts, playground, and small zoo. The zoo featured a bear den with a simulated pit surrounded by an iron fence, deer runs, and a monkey island. Parking spaces for the increasingly popular automobile were also included. The Works Progress Administration provided $100,000 under the supervision of L. J. Mills to make the suggested improvements. The zoo's superintendent had a budget of $1,200 a year for the zoo and earned a salary of $200 a month. He paid his zoo personnel 25¢ an hour for their 60-hour workweek. Oak Park housed two bears. Blacky the Bear was fed by sponsorship of M. J. Wendling. (Above, courtesy of City of Montgomery Engineering Department; below, courtesy of Montgomery Zoo.)

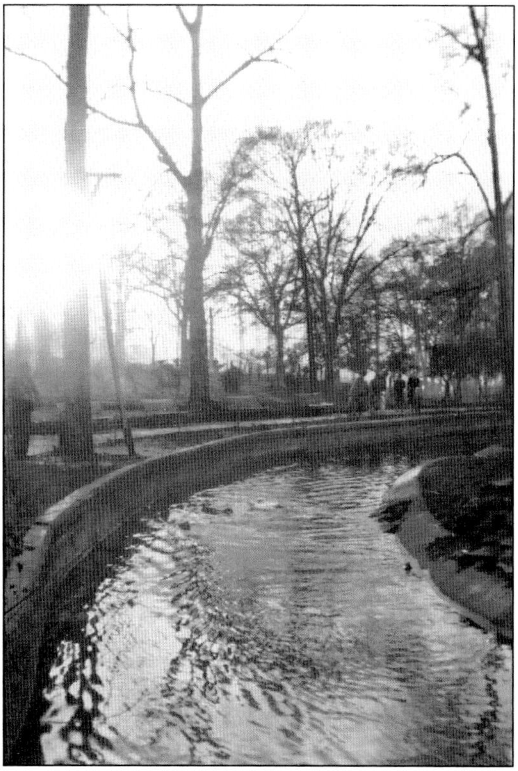

Various kinds of ducks and geese swam in the moat surrounding Monkey Island as it flowed under an arched rock bridge to the other animal areas. Spectators could view the monkeys' playful antics on the suspended playground from nearby walking paths. Monkey Island was unique for its time by providing the public an unobstructed view of the animals. A large enclosed expanse nearby confined the park's deer population. Visitors would often feed their picnic leftovers to the ducks. (Left, courtesy of Landmarks Foundation; below, courtesy of Montgomery Zoo.)

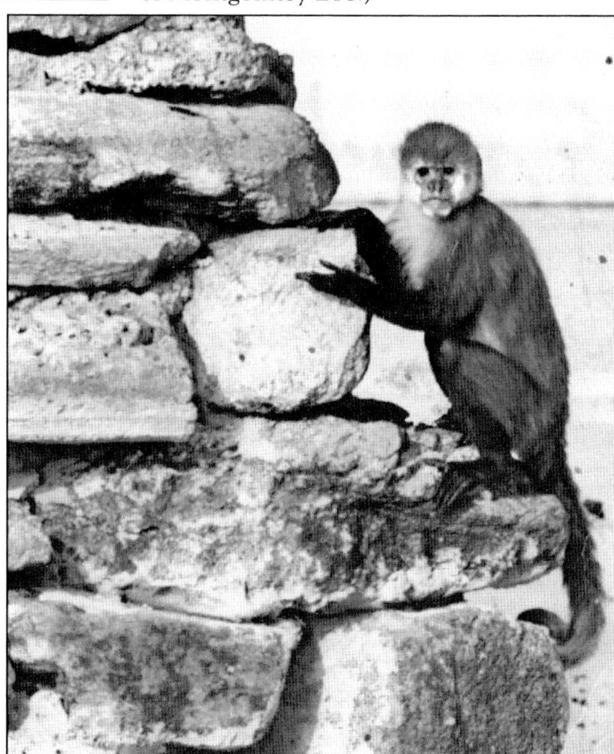

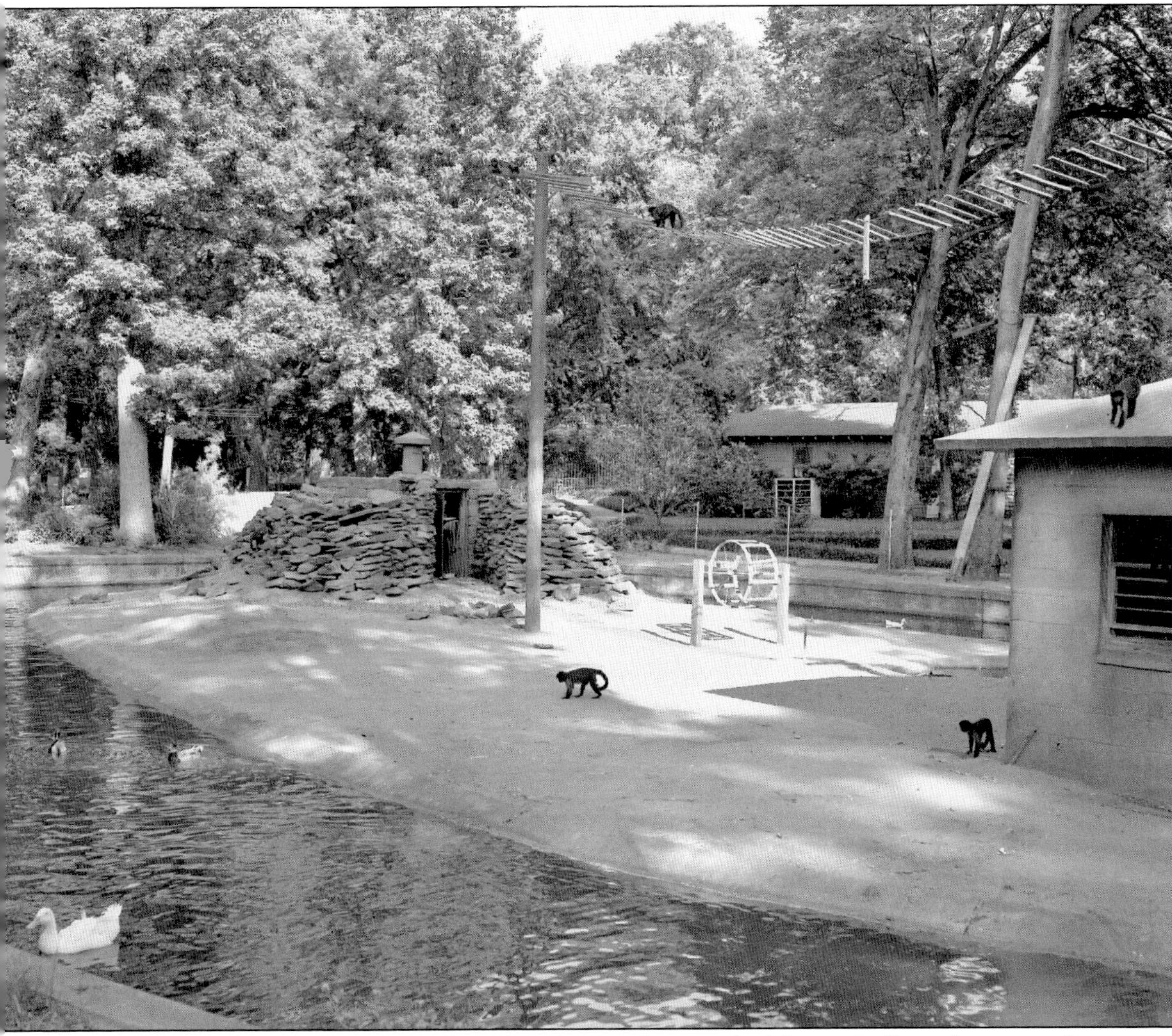

As Oak Park expanded, the zoo developed in the northern end of the park. Several species of monkeys (capuchins and woolley monkeys) lived on Monkey Island, an area created of stacked rocks and surrounded by a moat. The monkeys were provided with two indoor areas to escape the heat of summer sun. Children would flock to the park to watch the monkeys climb on their rocks and swing from the trees. From Monkey Island, the children could also view the other animals of Oak Park Zoo and listen to the parrots talk. Visitors can still see Monkey Island today at Oak Park without its inhabitants, or they can visit the Montgomery Zoo and see a similar exhibit. (Courtesy of Robert Fouts/Fouts Commercial Photography.)

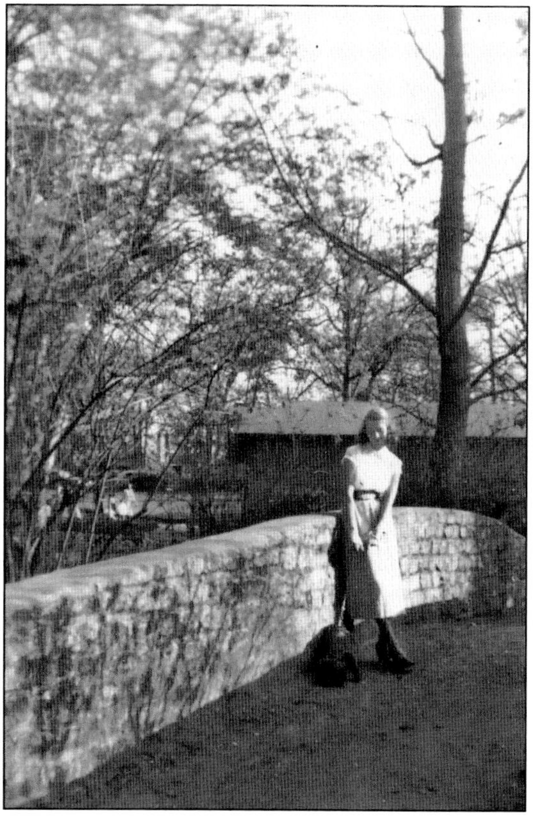

Betty Cargill stands on the rock bridge over the duck pond, the moat surrounding Monkey Island. By the early 1930s, Oak Park became an integral recreational outlet for the developing neighborhoods surrounding the area. Oak Park was now accessible by bus lines, which connected at Court Square downtown. (Courtesy of Cargill/McDuffee Collection.)

The earlier ornate wooden pavilion was demolished, and in 1937, a Tudor-style building was constructed of granite blocks. Paving bricks from the city streets were pulled up and used for the pavilion, making it a unique building. This new recreation building was situated at the main entrance to the park, which was then located at the intersection of Forest Avenue and Park Place. (Courtesy of Robert Fouts/Fouts Commercial Photography.)

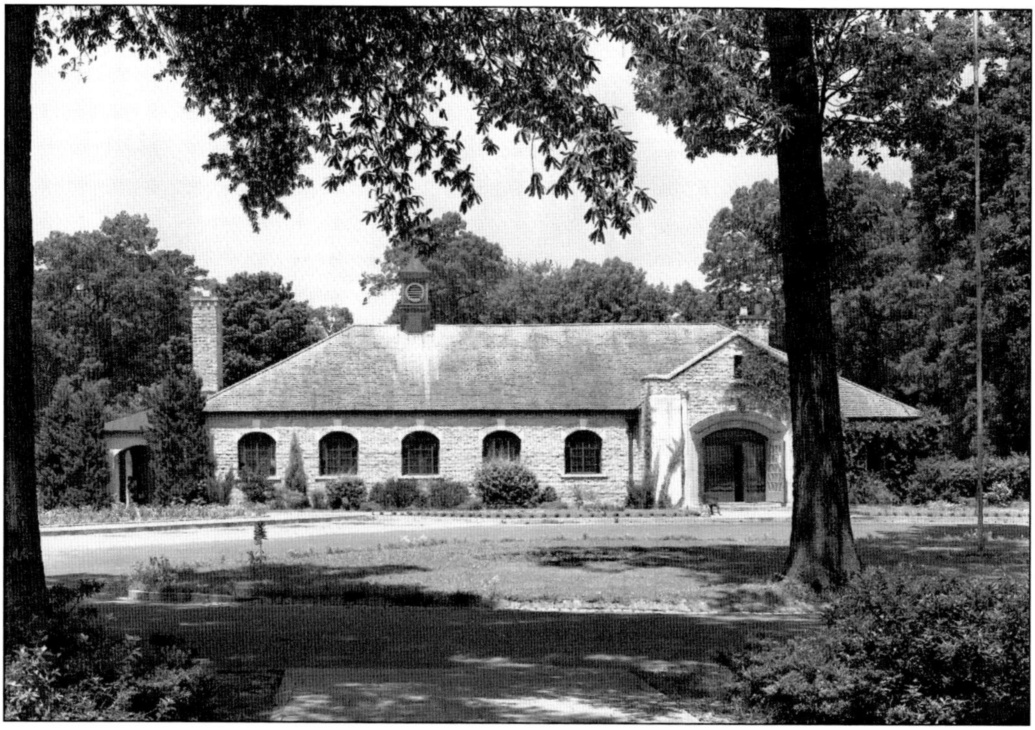

The colonnaded south entry to the main building opened to a formal garden accenting a reflecting pool. Behind the building, more formal flowerbeds trimmed the open vista to the swimming pool. The new pavilion was the last of the improvements made to Oak Park. (Courtesy of Cargill/McDuffee Collection.)

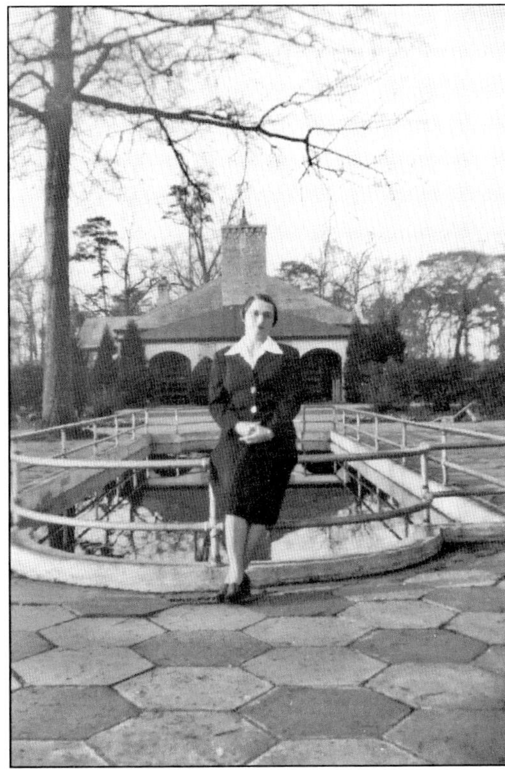

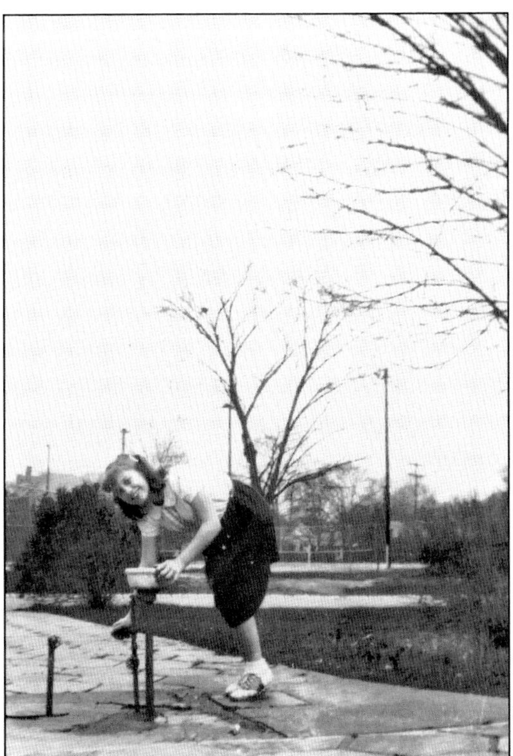

Oak Park had six clay tennis courts that were open for play during day or night. Pat Halstead poses for a picture while getting a drink from the wading pool fountain. The tennis courts can be seen in the background. (Courtesy of Cargill/McDuffee Collection.)

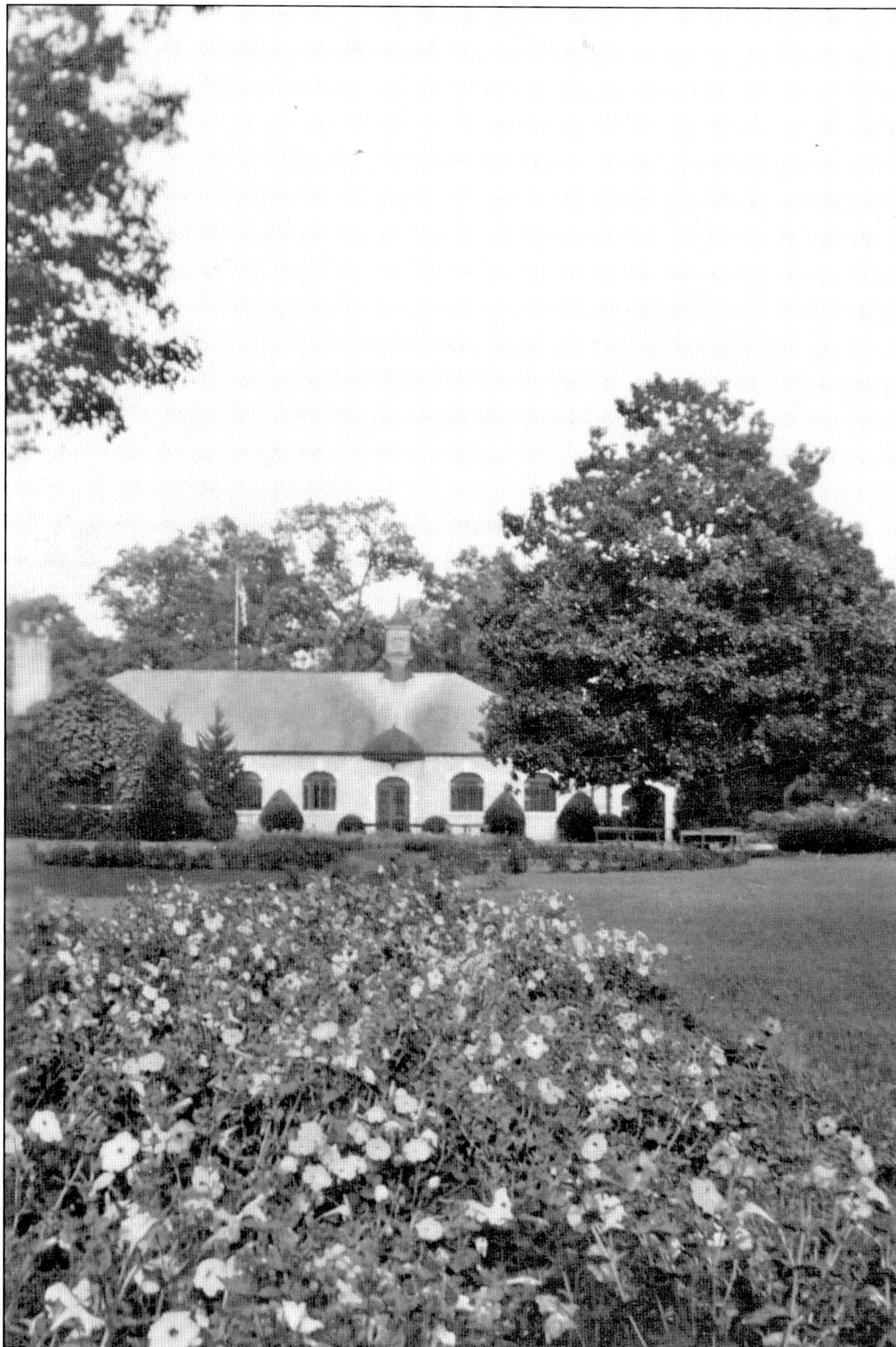

The emergence and availability of the automobile made driving to and through Oak Park a very popular pastime and social event. In order to protect the newly added flowers in the park, Section 50 of the Code of Ordinances of the City of Montgomery in 1938 stated that "No person shall gather, injure, or destroy any of the flowers, kill, shoot or throw at any birds, squirrels or animals in any park, or injure in any way, the trees, grass, shrubbery, or walk on the grass plots." (Courtesy of Cargill/McDuffee Collection.)

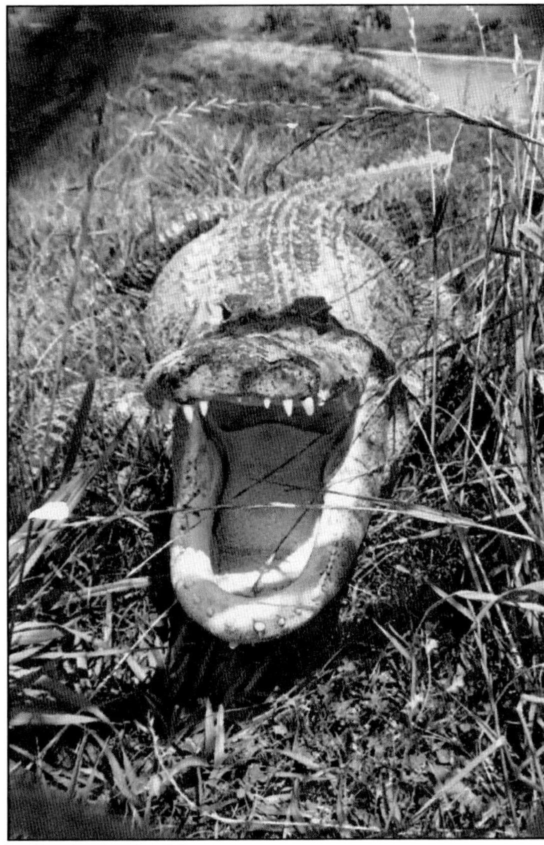

Animal diets of canned milk, fresh sweet bread, peanuts, and sweet potatoes were scrounged from the nearby A&P store. Horsemeat for the lions was purchased for 12¢ per pound; the cats apparently thrived on their diet of horsemeat and milk, for they both died of old age. The zoo at Oak Park contained a menagerie of animals. Some of the smaller animals were housed in cages that the children would walk by on their way to and from the swimming pool. (Courtesy of Montgomery Zoo.)

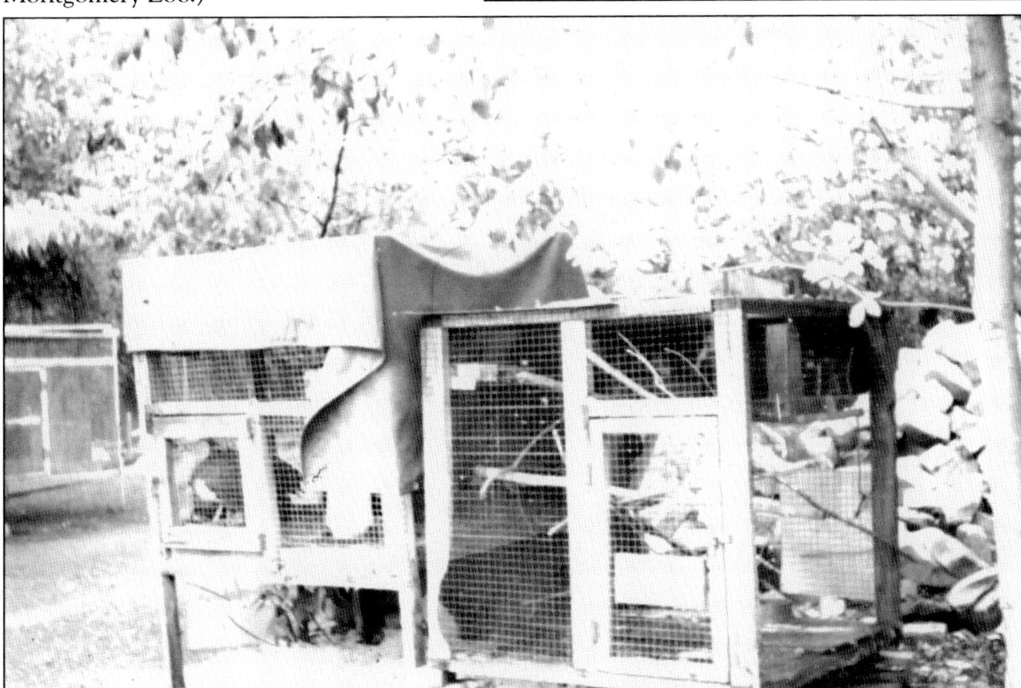

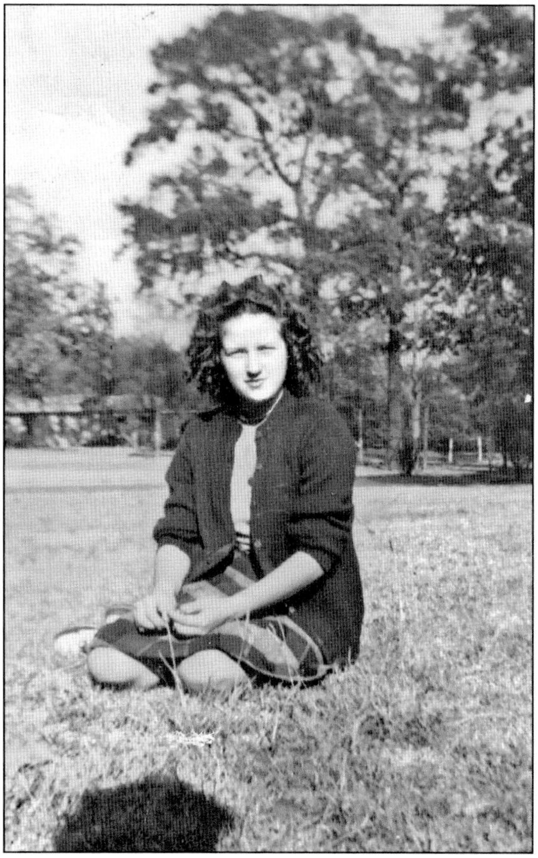

Posing in the park for the photograph at left, with the bathhouse and swimming pool in the background, is Martha Hester Green in 1940. The bathhouse provided changing rooms for children who came to the swimming pool in Oak Park. Harold Green (left) and Martha Hester pose with friend, Leslie (right), in front of the lion cage. There is a tale of a man getting his arm bitten off by the lion told in two ways. Some people believe that one night a drunken man approached the lion cage and attempted to feed the lion a beer. The lion grabbed a hold of his arm, tearing it from his body. The other version involves a veterinarian, whose office was across from the park on Mulberry and East Fourth Street, being badly mauled by the lion while trying to feed him. (Courtesy of Martha and Howard Green.)

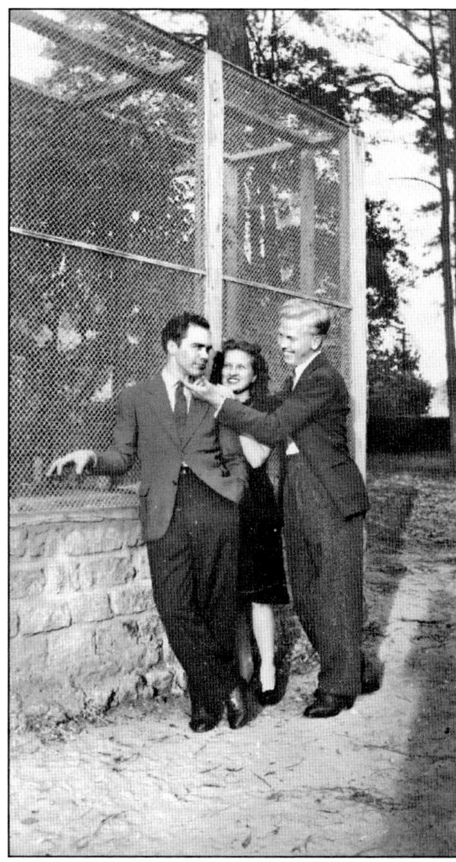

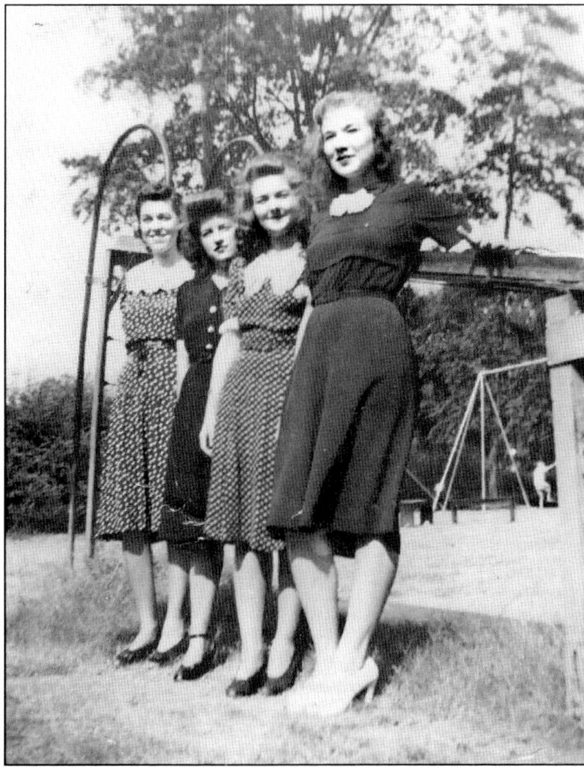

The playground area of the park featured swings, a sliding board, and a merry-go-round. The sliding board was a unique type of slide that had rollers. Even older children and adults enjoyed the playground. Pictured at right from left to right are Peggy Cauthen, Martha Hester Green, June Dillard, and Pat Cauthen, who posed for a photograph in 1941. Below are their male counterparts, from left to right: Lee Dillard, Fletcher Cauthen, Freddie Cauthen, and Harold Green. (Courtesy of Martha and Howard Green.)

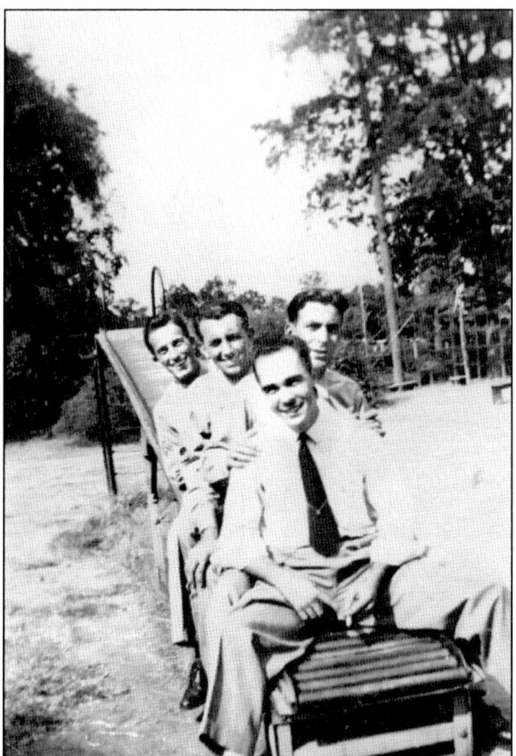

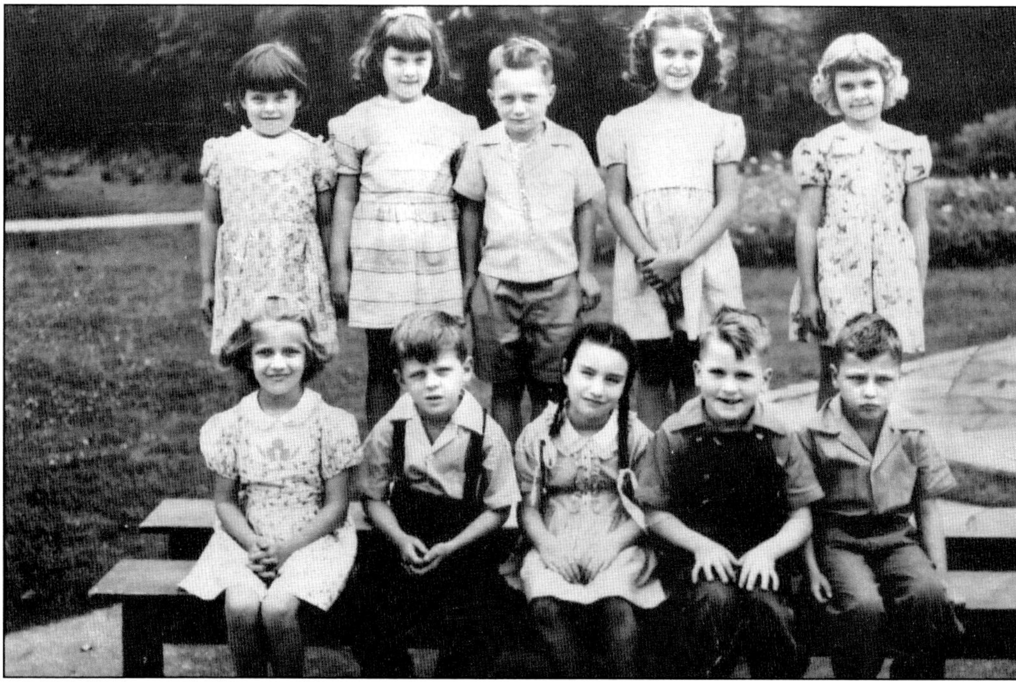

Kindergarten classes were held in the pavilion at Oak Park and in 1946 were taught by Mrs. Knox. After kindergarten, children in the area would attend Forest Avenue Elementary School. Class photographs were taken of the children posing in front of their classrooms. The children were often taken to the park during and after school. The teachers were very strict about them wearing shoes to the park—even the students who would attend school barefoot—because of the polio scares. (Above, courtesy of June Thompson; below, courtesy of V. Norman Dennis.)

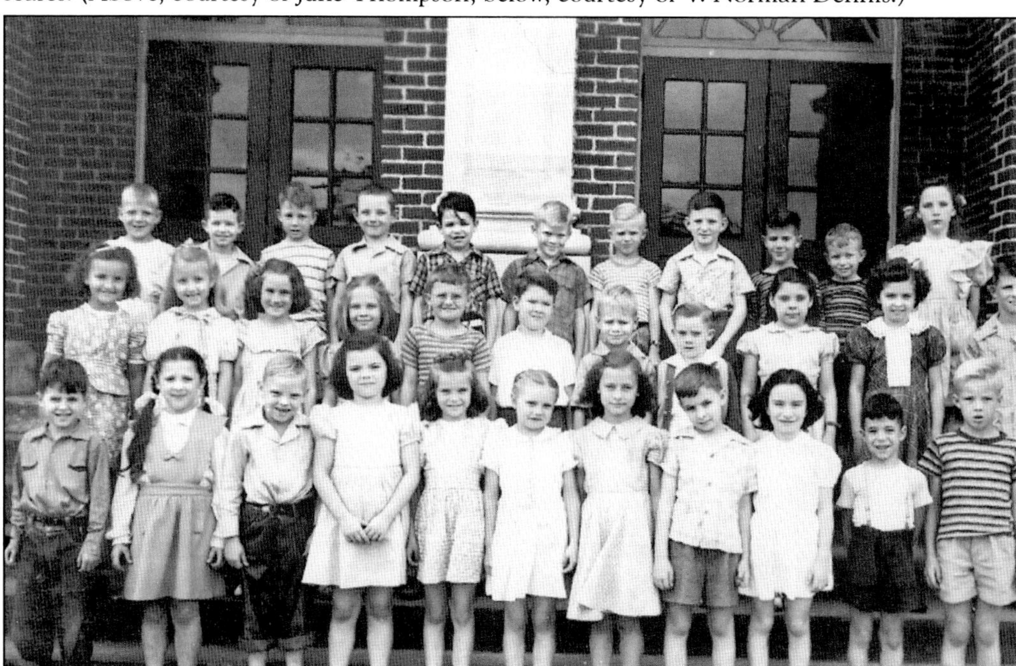

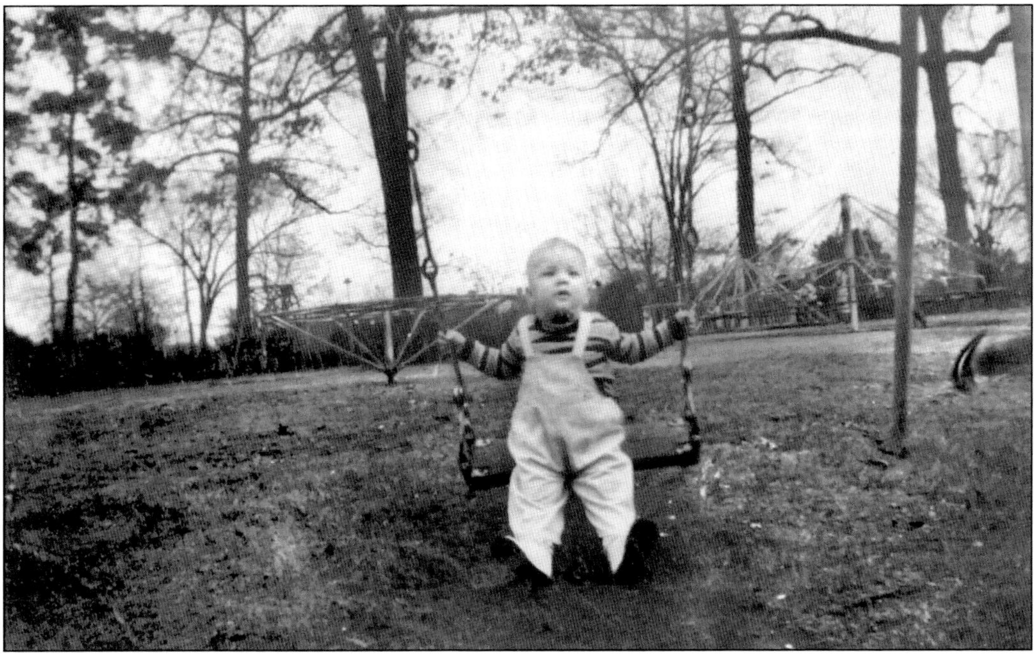

Little Johnny Green enjoyed his ride on the swing at Oak Park in 1949. The playground not only featured swings and the unique sliding board, but it also had a merry-go-round. Children would run in circles while holding onto the handrails of the merry-go-round and then jump on for an exhilarating ride. (Courtesy of Martha and Harold Green.)

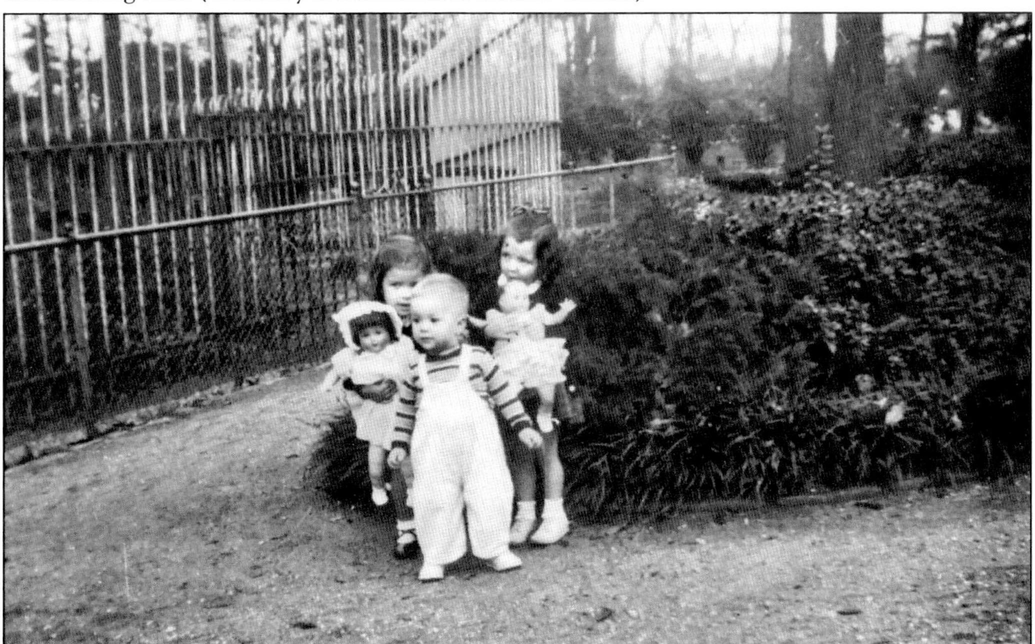

Near Monkey Island were other larger monkeys housed in a fenced enclosure. The zoo provided a variety of animals for children and adults to enjoy. Gathered around the monkey cage, Sandra (left) and Johnny Green (center) and friend, Debby Webster Thornton, pose for a photograph in 1949. (Courtesy of Martha and Howard Green.)

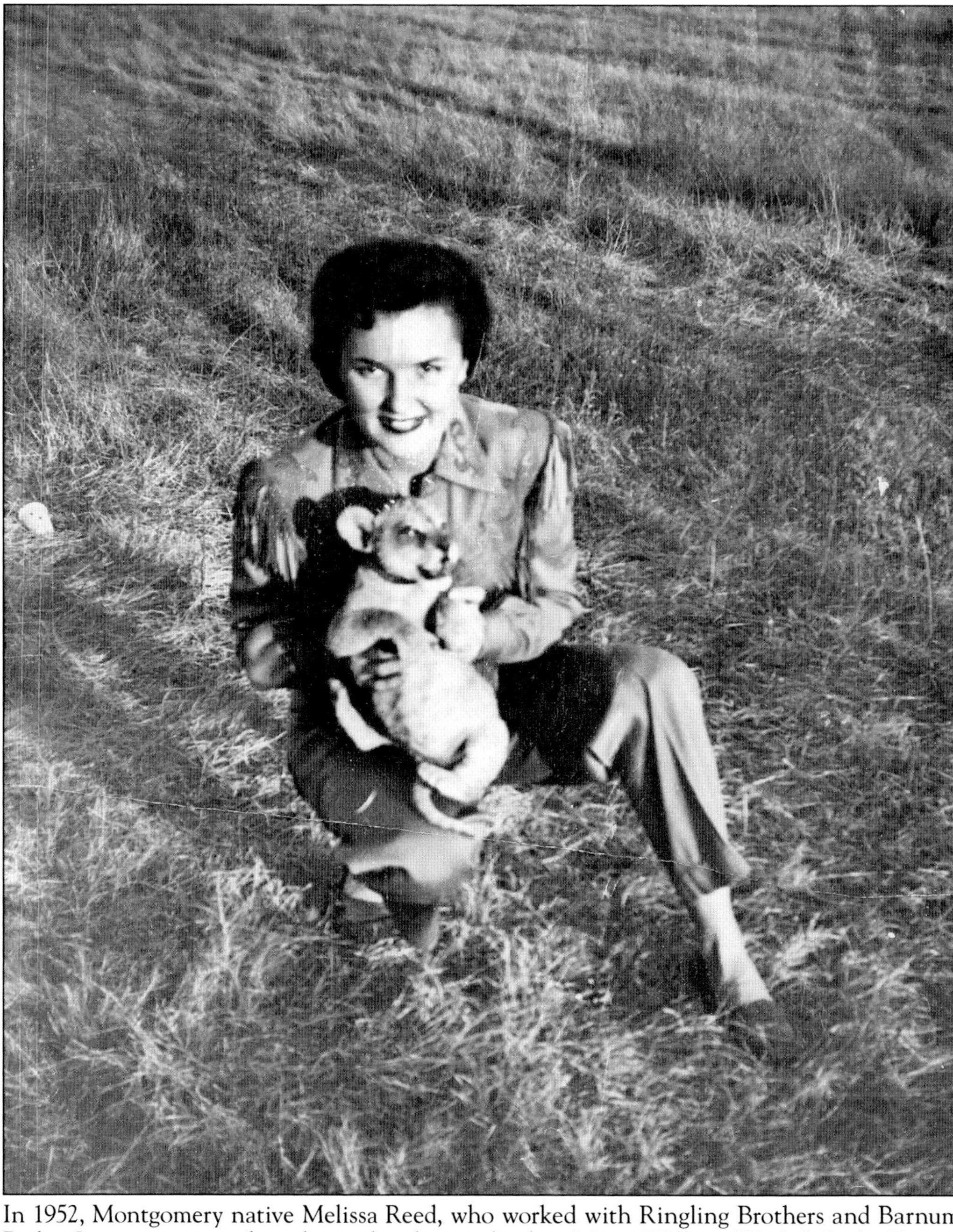

In 1952, Montgomery native Melissa Reed, who worked with Ringling Brothers and Barnum Bailey Circus, sent a newborn lion cub to her uncle, then-Mayor William "Tacky" Gayle. Known as Molasses, the lion cub lived with the mayor at his home on Gilmer Avenue. Molasses quickly became a celebrity around town and rode with the mayor in his car to city hall daily. However, when she outgrew her Garden District home, the mayor presented Molasses to the Oak Park Zoo. When the park was closed in 1959, Molasses found a new home at the Birmingham Zoo. (Courtesy of Melissa Reed.)

Mayor W. A. Gayle raised Molasses in his home. She would ride with him in his car to and from work every day. When she finally was too big to take care of, Molasses found Oak Park her home. Another lion that was purchased from a zoo in Arizona later joined her. (Courtesy of Montgomery Zoo.)

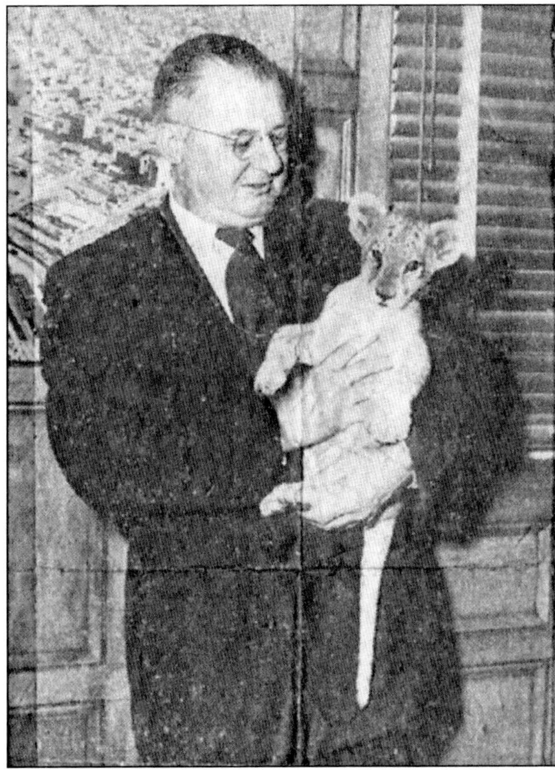

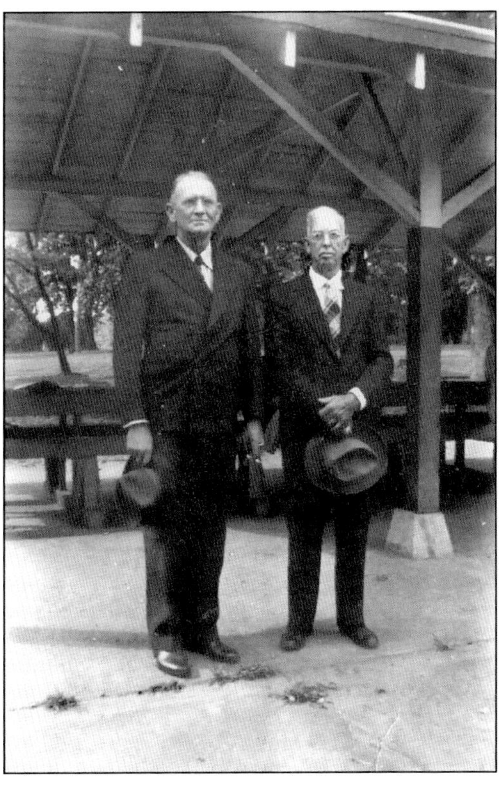

Frank Smith (left) and his nephew, James Guy, are photographed during a family reunion in Oak Park under one of the pavilions provided for outdoor activities. Smith (born in 1890) traveled with his family from Texas to Montgomery for this special event. Guy (born in 1882) and his family moved to Montgomery in the late 1930s. (Courtesy of June Thompson.)

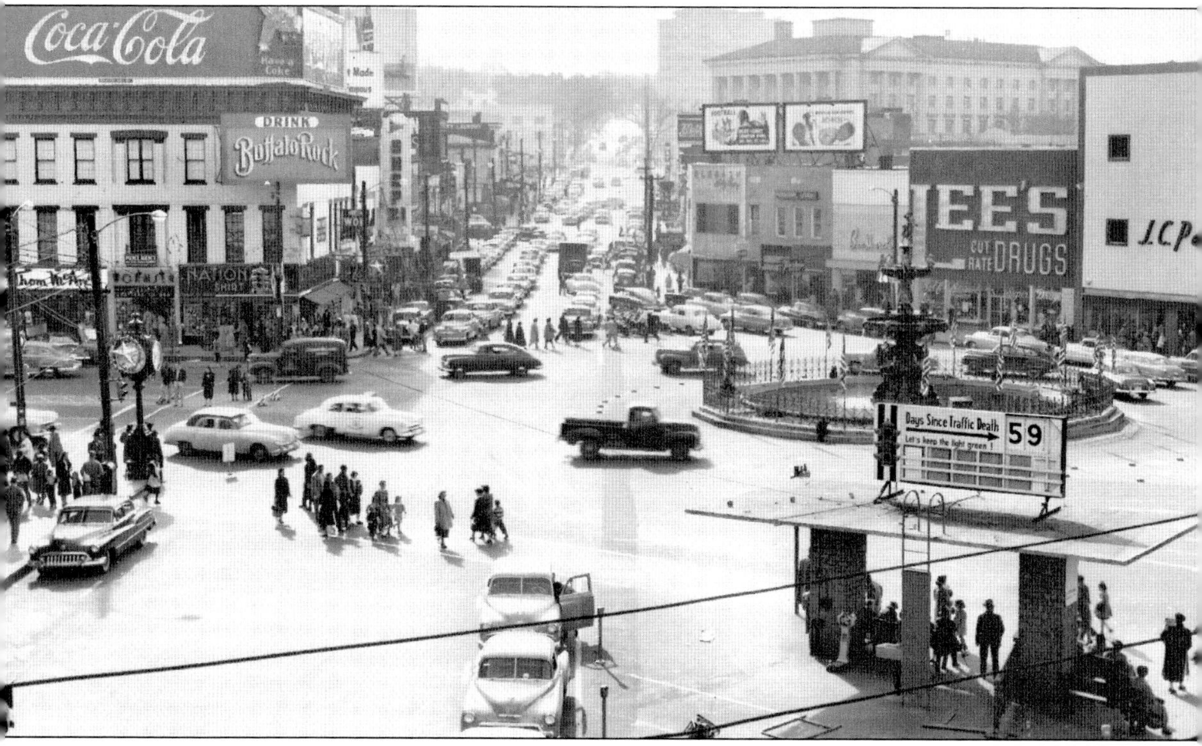

Rosa Parks, on December 1, 1955, refused to give up her bus seat to a white man, leading to her arrest and a boycott of city buses. The Montgomery bus boycott of 1955 lasted 11 months, forcing the city to desegregate its transit system. A federal district court ruled that Alabama and Montgomery's racial segregation laws for public buses were unconstitutional on June 4, 1956 (*Browder v. Gayle*). Appeals pushed the case to the Supreme Court while maintaining segregation on buses. Finally, on November 13, 1956, the Supreme Court upheld the federal district court's decision that segregation on public transit was unconstitutional. A city ordinance on December 20, 1956, ended the Montgomery bus boycott by allowing blacks to sit anywhere on public buses. (Courtesy of John Scott.)

In 1956, the Army Signal Corps disbanded the pigeon breeding center at Fort Monmouth, New Jersey, which was home to at least 18 pigeons that served in combat. Lead by the Blue and Gray Association, Mayor Gayle and Paul Fuller, general manager of the chamber of commerce, petitioned the army for any pigeons that needed homes. In April 1957, Julius Caesar (above) and Flipper (below) were presented to Oak Park with a special ceremony. A blue-check-splash male pigeon, Julius Caesar is credited with a total of 44 combat missions during World War II, including a 300-mile flight crossing the Mediterranean to deliver an important message to American troops in Tunisia. He is among the top six pigeon heroes to be in over 40 wartime missions. Flipper, a blue-check female pigeon, distinguished herself as a World War II heroine. She carried more than 20 combat missions, including several that followed parachute landings. (Courtesy of U.S. Army Communications-Electronics Museum.)

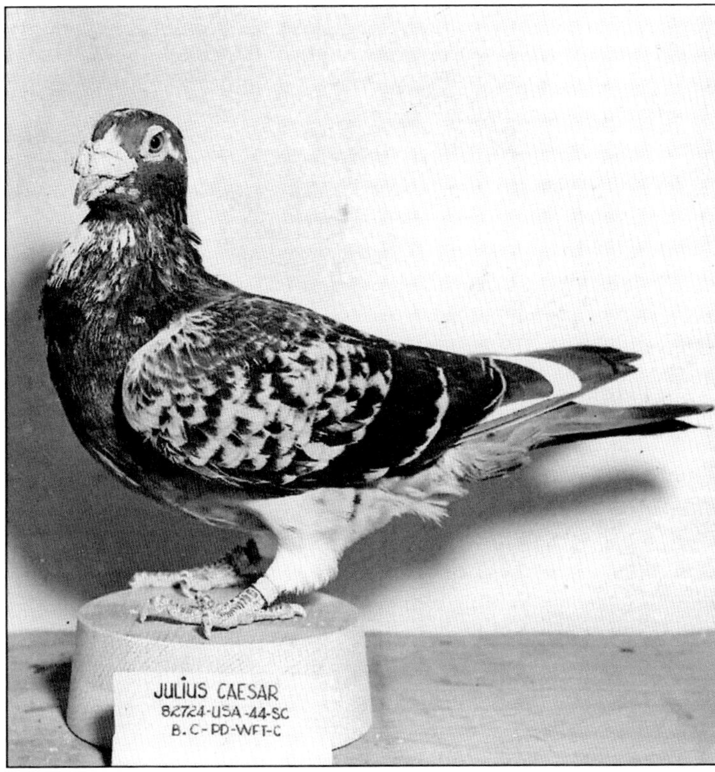

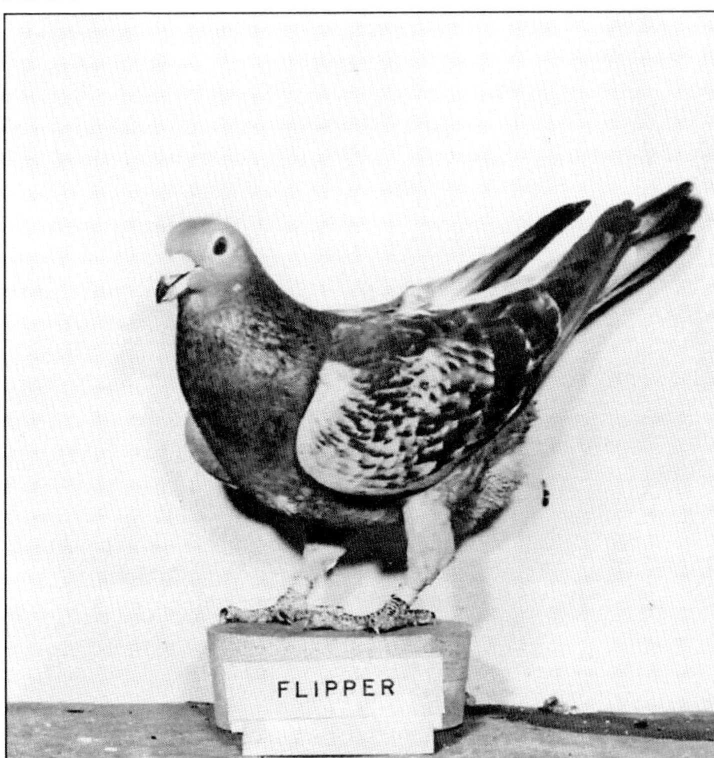

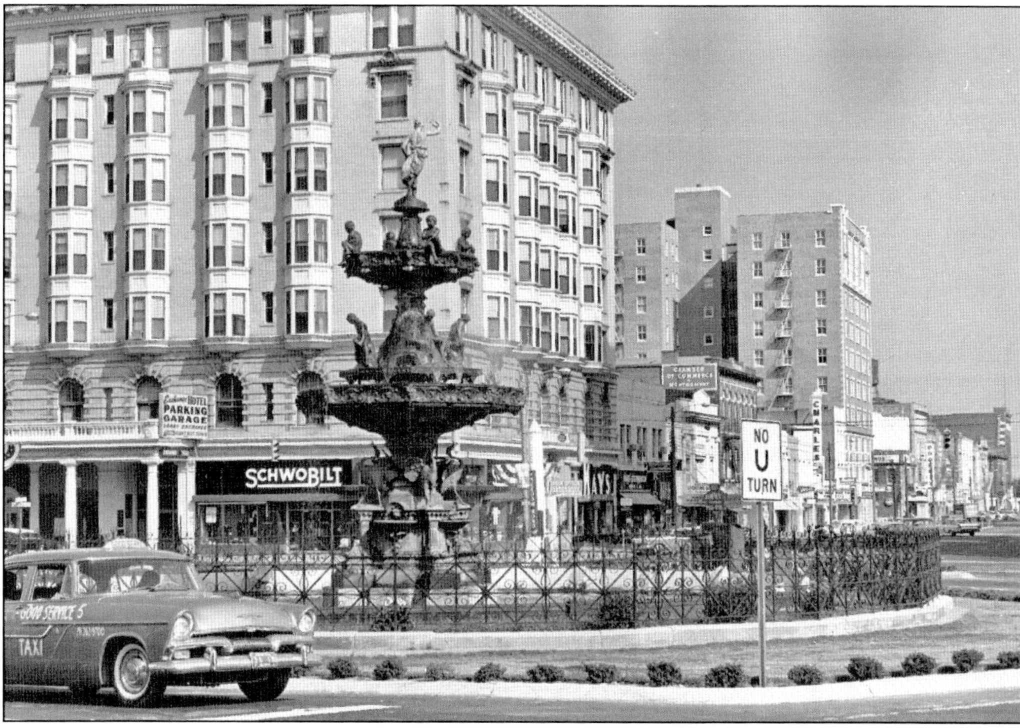

In September 1958, a group of black citizens of Montgomery filed a class-action lawsuit to desegregate the public parks of Montgomery. Federal court judge Frank Johnson ruled that city's recreational facilities policy was unconstitutional in 1959. Within four months of the ruling, city officials decided that rather than desegregating the pubic parks and swimming pools, they would close all of the parks. At Oak Park, the swimming and wading pools were drained; the lions, bear, alligators, peacocks, monkeys, and deer were either given away or sold. After almost 60 years as a local landmark, the gates of Oak Park were closed in December 31, 1959. (Above, courtesy of Landmarks Foundation; below, courtesy of Cargill/McDuffee Collection.)

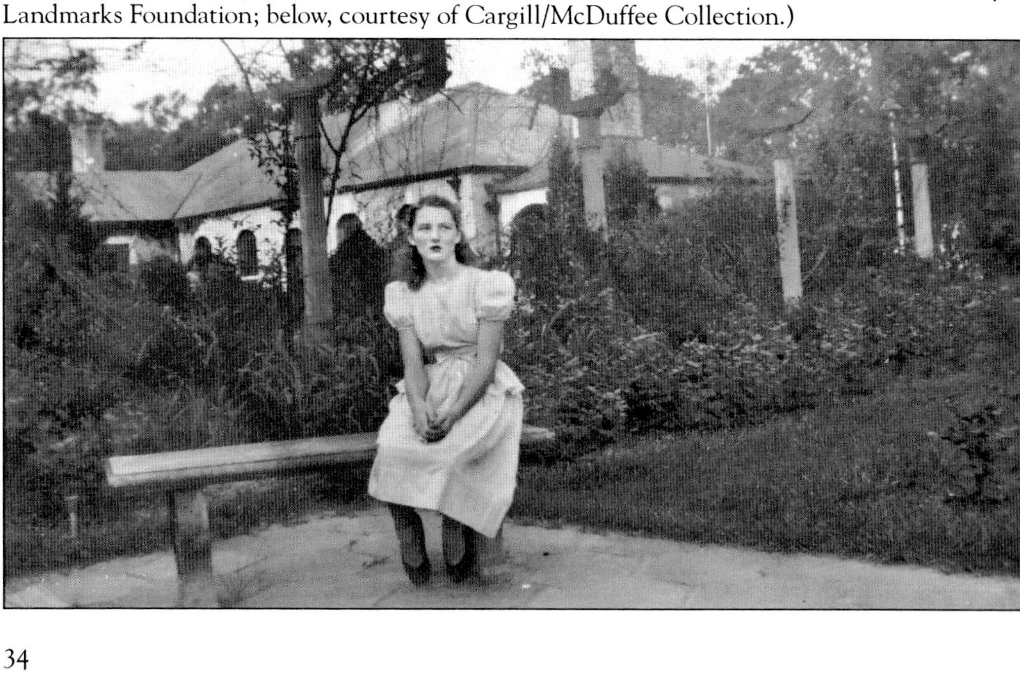

Two

Montgomery Children's Zoo 1967–1974

In 1967, plans were approved for a 34-acre recreational park in north Montgomery, including acreage for a small zoo. The site selection was mostly a treeless tract adjacent to the Kilby Prison farmlands; much of the site, which at one time had been used as a racetrack, was low, swampy, and required extensive filling. Funds were appropriated by the city for the initial zoo construction, consisting of a zoo office/concession, a red barn, a flight cage, and a monkey island. Completed in the spring of 1971, these first facilities were aesthetically pleasing, but the lack of design input from zoo professionals left much to be desired.

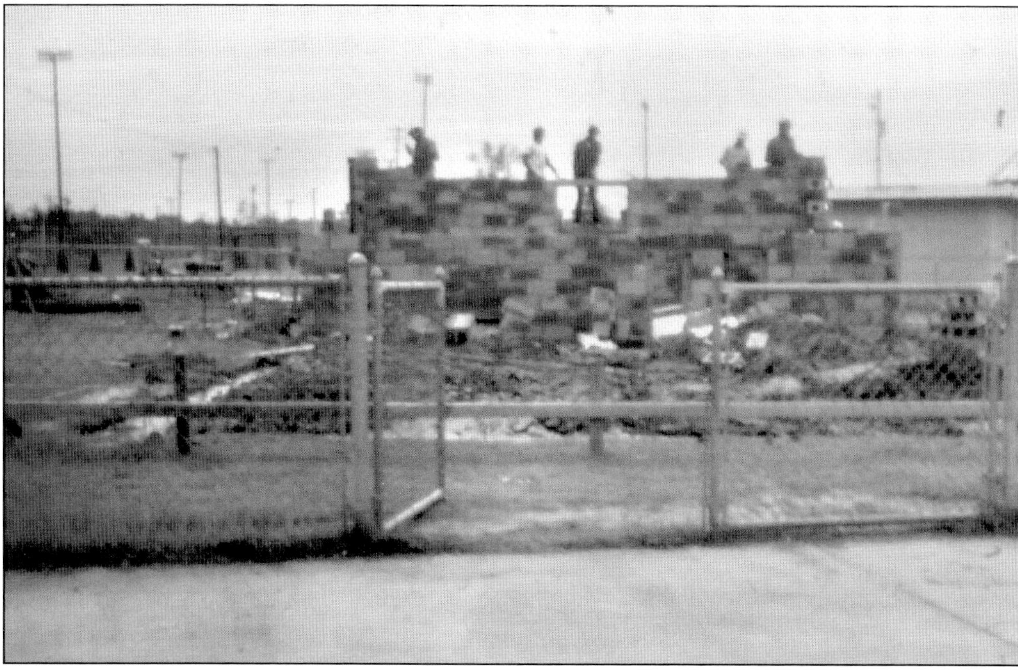

After the closing of the Oak Park Zoo, Montgomery went for seven years without a local zoo. Finally in 1967, the city approved plans for a small 34-acre zoo and recreational complex in north Montgomery. In October 1971, the city's Parks and Recreation Board voted to name the newly created zoo the Montgomery Children's Zoo. The zoo, which was being stocked and readied for opening, featured small and domestic animals for youngsters to pet. A larger zoo with wild jungle animals was being planned at Lagoon Park, adjacent to Gunter Air Force Base. (Courtesy of Montgomery Zoo.)

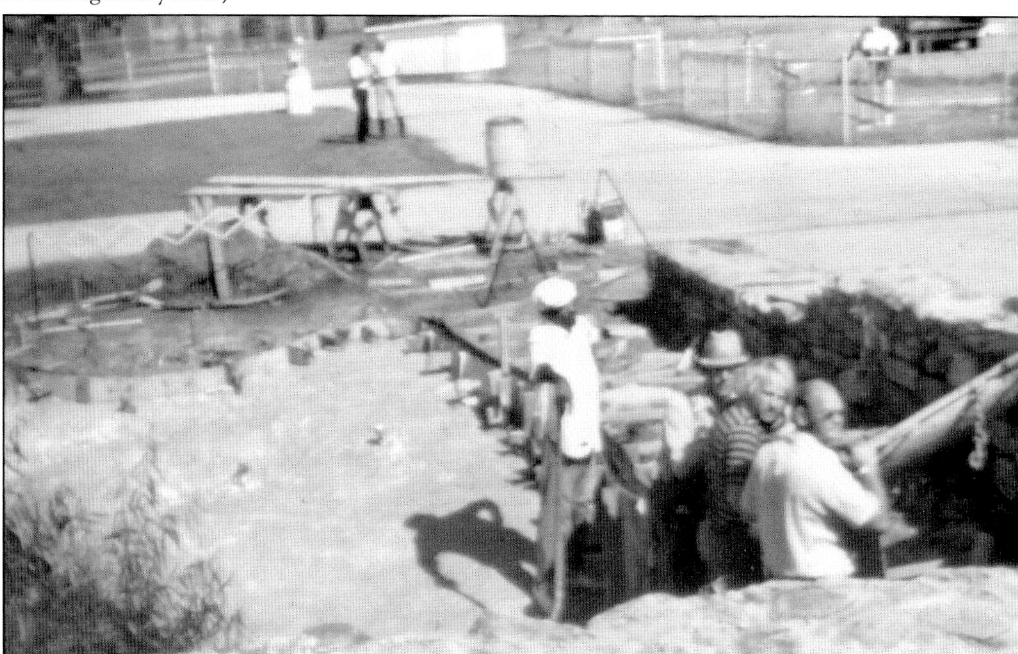

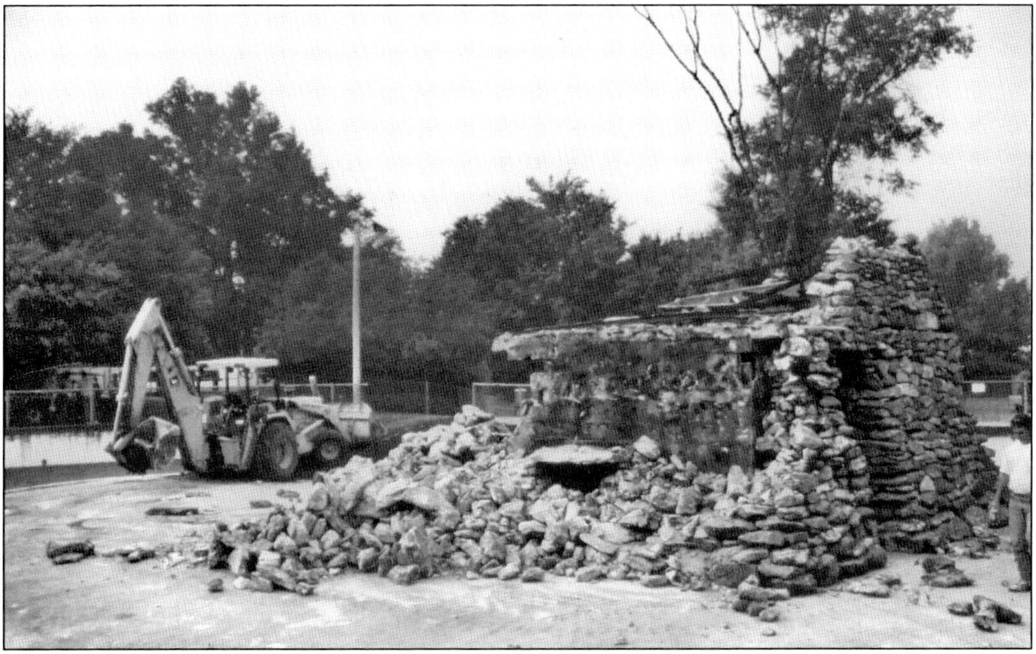

Like Oak Park Zoo, the Montgomery Children's Zoo planned to feature a monkey island, exhibiting several species of monkeys including capuchins. The original Monkey Island influenced the plans of the new habitat. Surrounded by a moat, the monkeys could play on their rock-covered house and hang from the trees and bridges. (Courtesy of Montgomery Zoo.)

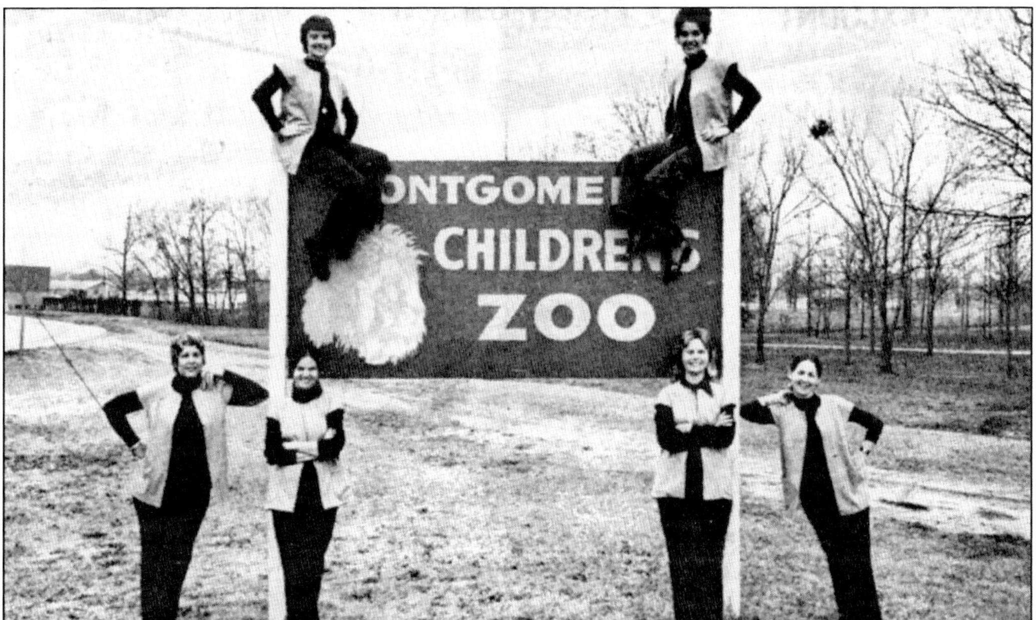

Montgomery's new tennis complex opened in August 1971, but the opening of the new zoo was delayed. The city decided to find a director before the grand opening. The new director had to stock and equip the zoo before opening. Members of the Capitol Junior Women's Club served as guides at the Montgomery Children's Zoo. (Courtesy of Montgomery Zoo.)

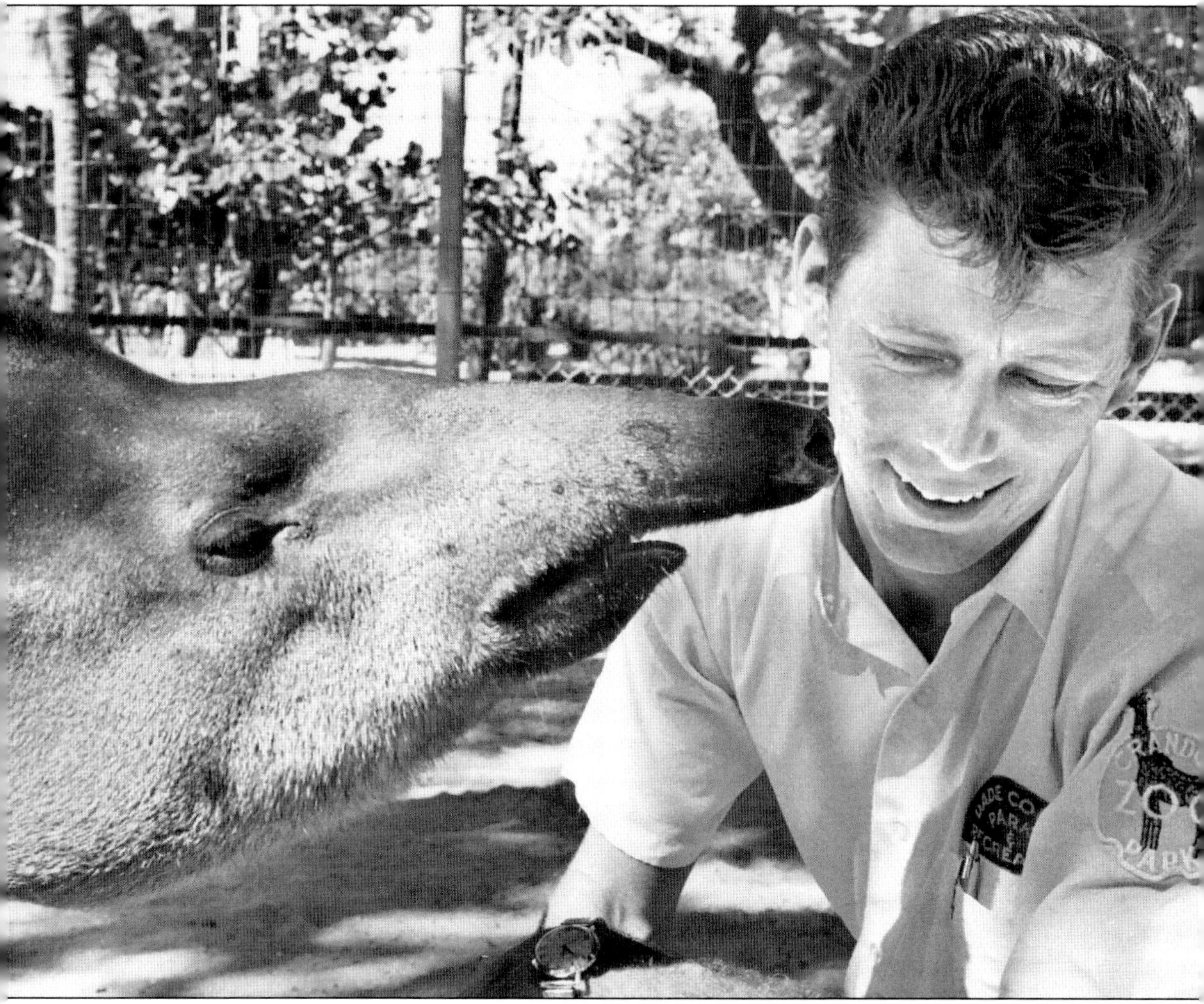

Four years after beginning to plan and build a new zoo for Montgomery, Jim Amon was hired as the new zoo director on September 16, 1971. Originally from Montgomery, Amon had worked with the Crandon Park Zoo in Miami, Florida, for the past 10 years. Starting as a zookeeper, he was promoted to curator of mammals and than publicity director. He also made weekly appearances on WTVJ's children's program, *Popeye Playhouse*. Tragedy struck the new Montgomery facility even before it opened. Jim Amon died in an auto accident just 37 days after becoming the director. (Courtesy of the *Miami Herald*.)

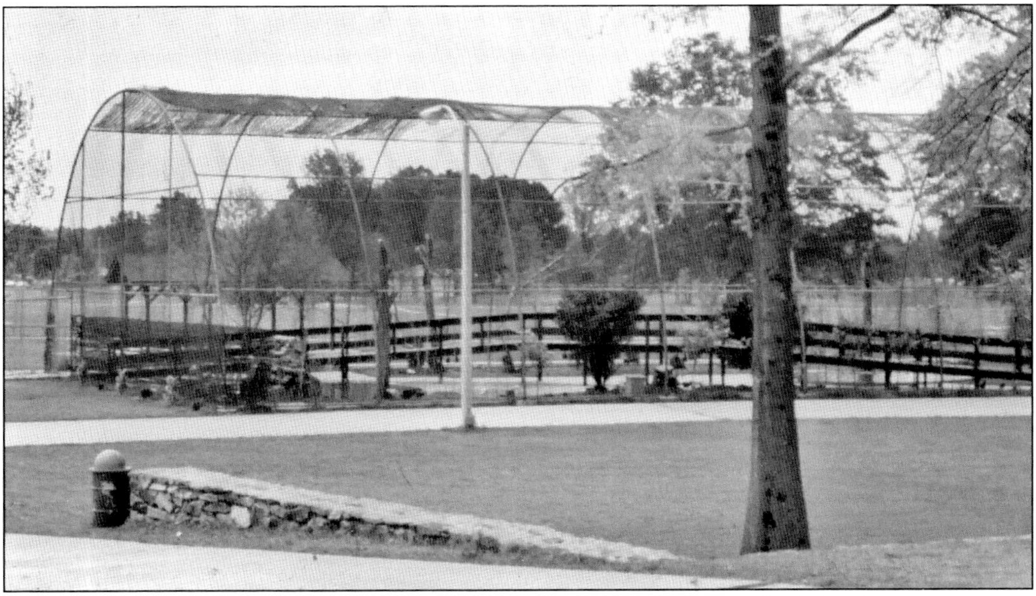

The new zoo had many design flaws even before it opened, because of a lack of input from zoo professionals. The flight cage floor was built of a solid expanse of concrete, the access to Monkey Island was by means of an overly sophisticated but minimally functional hidden bridge, and the sewage system proved quite inadequate. All of these problems had to be fixed before the zoo could open to the public. (Courtesy of Montgomery Zoo.)

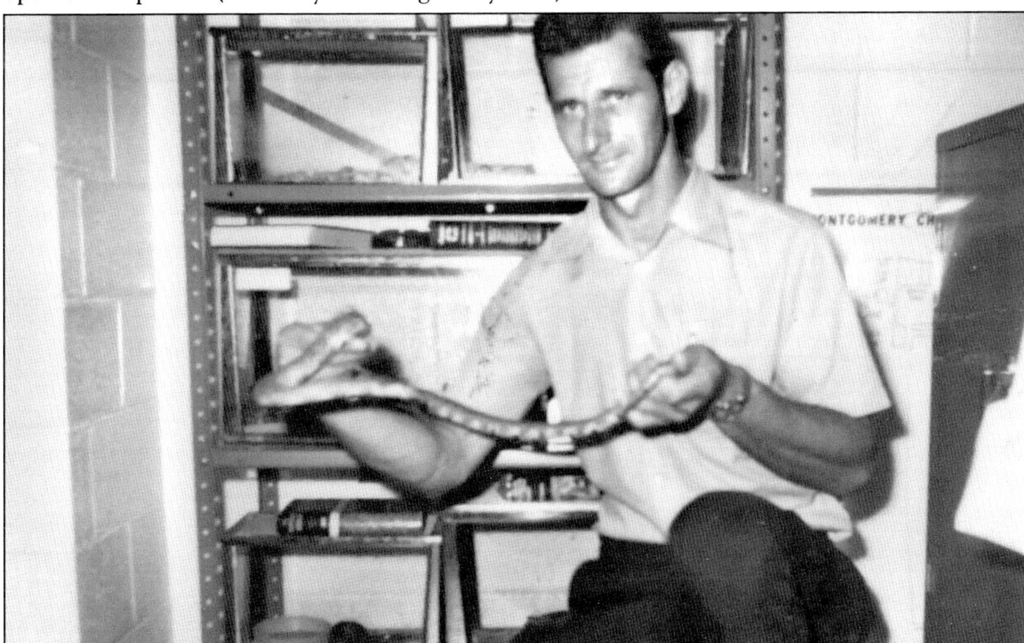

When the zoo hired Amon as new director, they also hired their first zookeeper. Paul Foster began his lifelong career with the Montgomery Children's Zoo in September 1971. After Jim Amon's passing, he was appointed as acting director. He held this position until March 1972, when the zoo hired a new director. Foster would again hold an interim director position in later years. (Courtesy of Montgomery Zoo.)

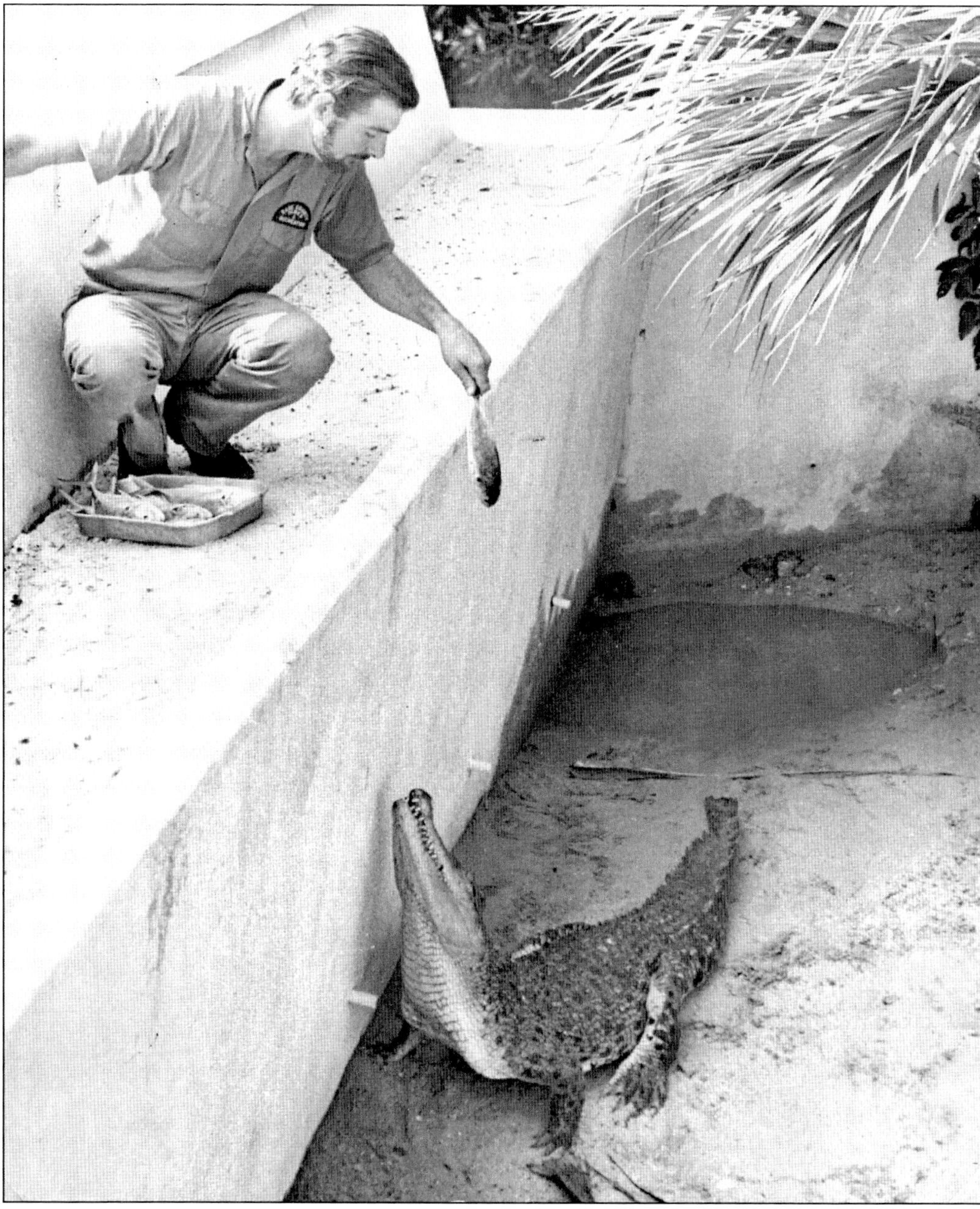

John Heidger, curator at Crandon Park Zoo, was hired as the second director of the Montgomery Children's Zoo in March 1972. Heidger's career did not begin with animals. He spent six years in the army and then went to work for a private investigation agency. When interviewed for the job at the Crandon Park Zoo, he said that he never expected to work in a zoo but applied after seeing an advertisement in the newspaper. He fell in love immediately with the work he was doing and was promoted quickly within the zoo. Heidger's main mission was to bring more exotic animals to the zoo and to develop a volunteer staff to lead educational tours. He did not want to be limited as a petting zoo, as the name of the zoo suggested. (Courtesy of Joe Migon.)

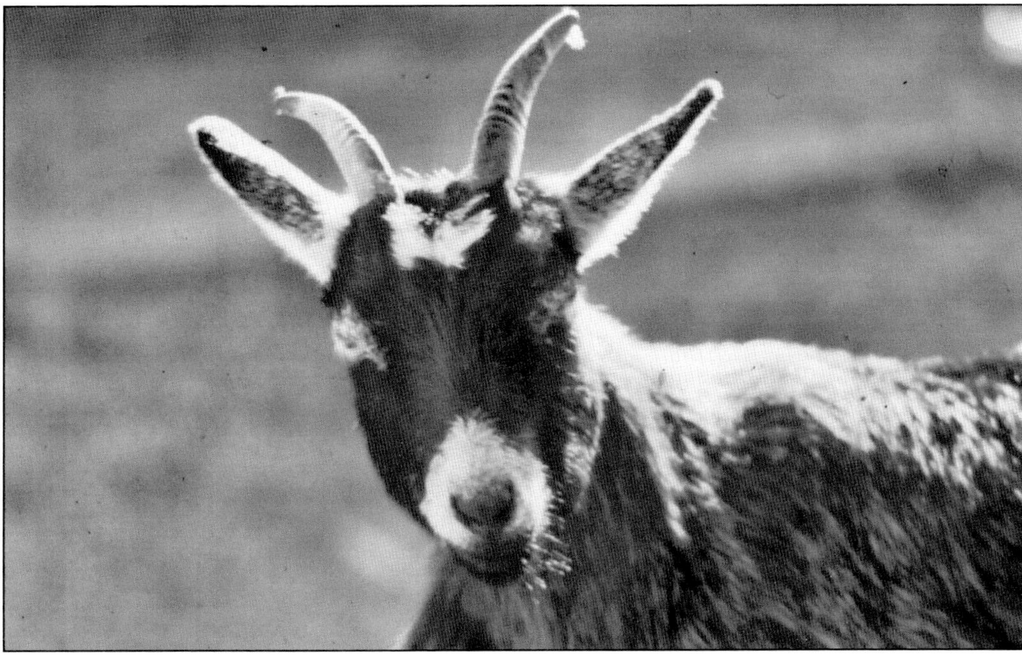

The Montgomery Children's Zoo was designed to be a petting zoo. Children and adults were able to interact with most of the animals either at the petting zoo or at the flight cage. Domesticated animals like goats and sheep were obtained to fill the zoo. (Courtesy of Montgomery Zoo.)

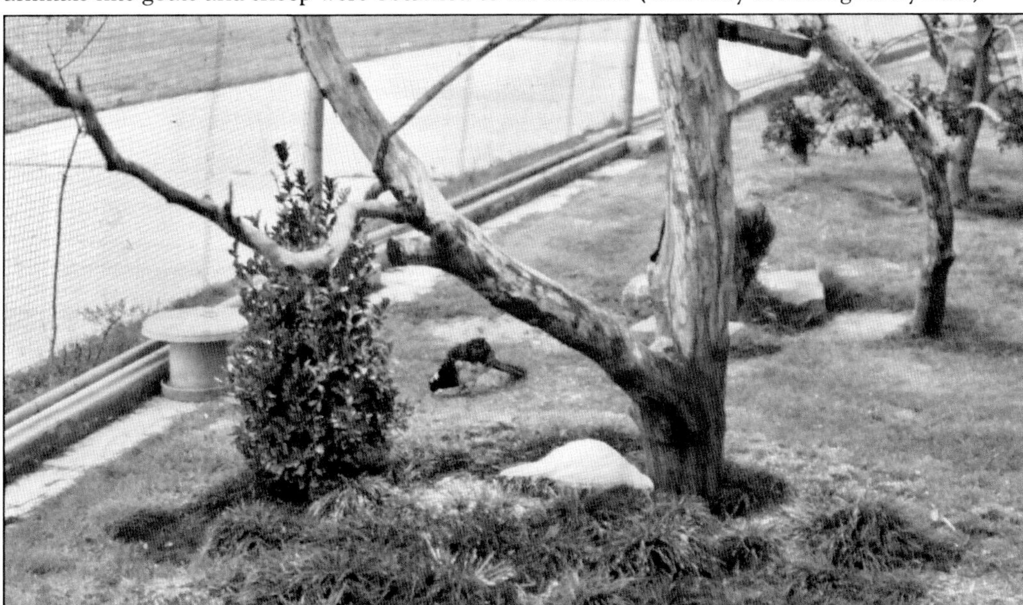

The flight cage was remodeled before opening to remove the all-concrete floor. A wooden bridge crossed the flight cage, where many of the zoo's birds live. Birds such as peacocks, guineas, and chickens roamed free. Among those inhabiting the flight cage were flamingos, gulls, ducks, vultures, a toucan, pheasants, and lovebirds. Nested in the shrubs, rarely noticed by visitors, was an iguana. (Courtesy of Montgomery Zoo.)

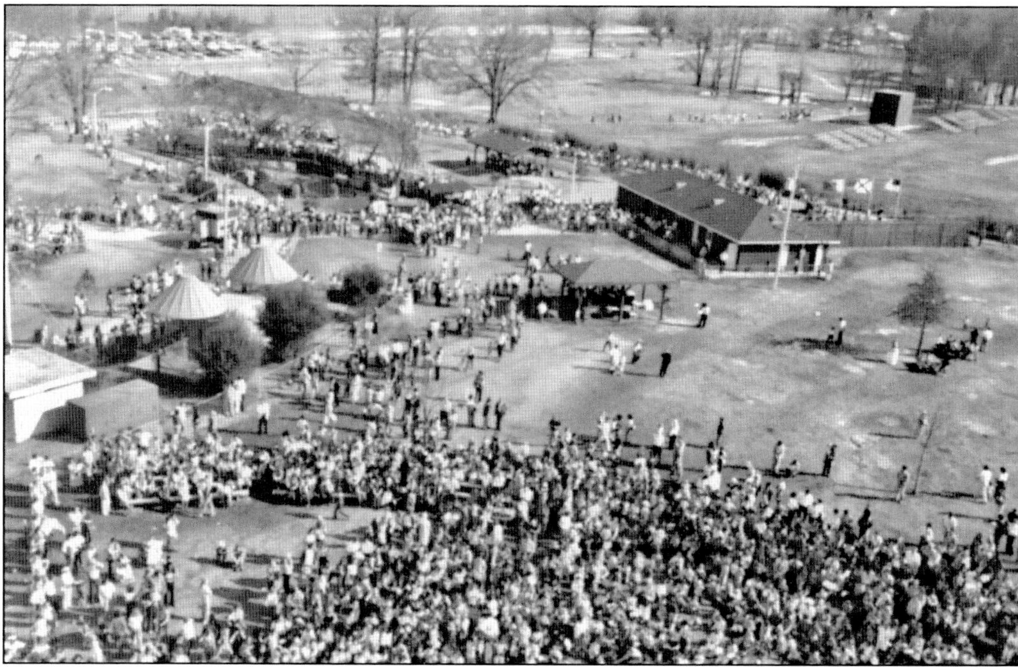

After many starts and stops, the grand opening finally arrived. The Montgomery Children's Zoo officially opened on May 6, 1972. Mayor Jim Robinson participated in the ribbon-cutting ceremony. After four years of planning and work, the children's zoo was a reality. With an admissions price of 50¢, the zoo welcomed its first guests, totaling 3,800. (Courtesy of Montgomery Zoo.)

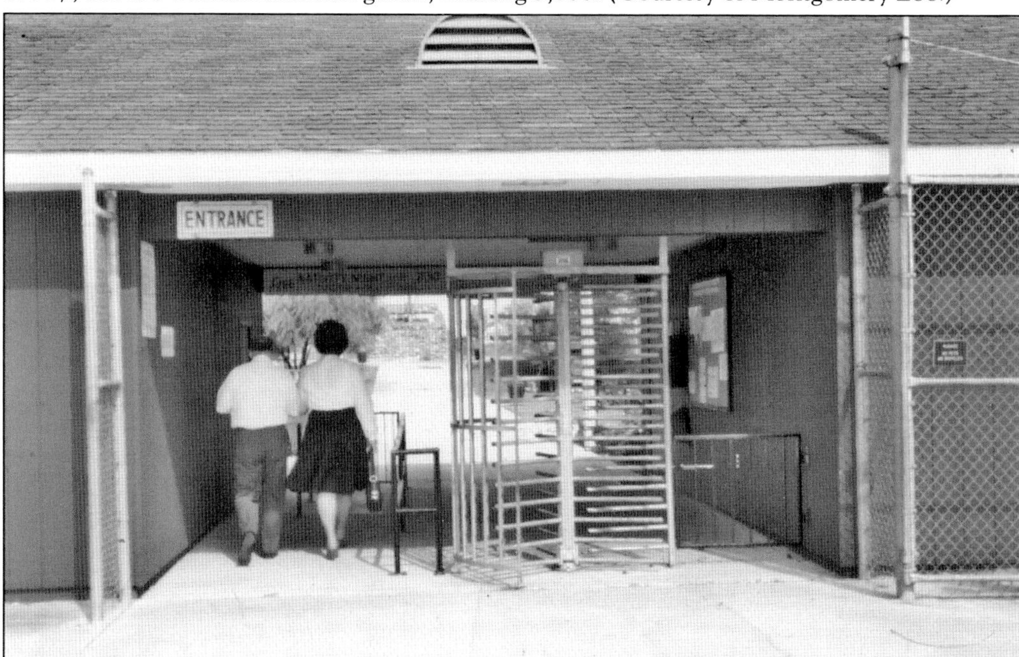

The zoo had areas for picnicking, including a picnic pavilion right inside the front entrance. There was also a concession stand. The zoo was open from 9:30 a.m. to 4:30 p.m. seven days a week. Admission was 50¢ for adults and free for children. (Courtesy of Montgomery Zoo.)

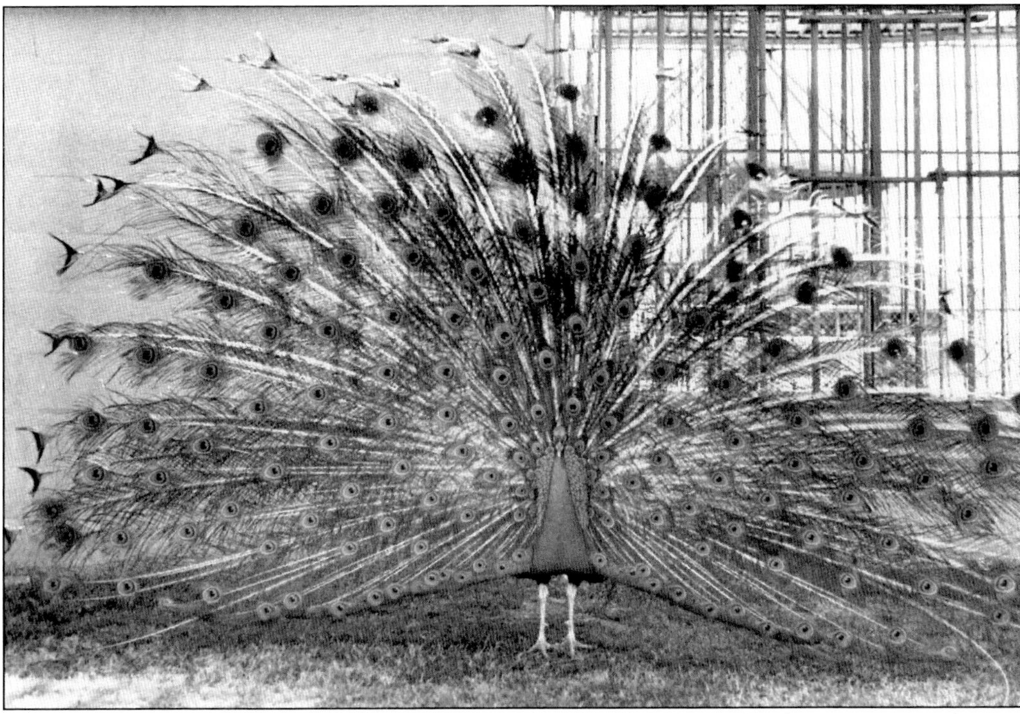

The first inhabitants of the Montgomery Children's Zoo included an 8-foot boa constrictor donated by the Riverside Speedway Racetrack in Montgomery. Other animals were flown in from zoos around the country, such as Crandon Park Zoo, where both of the zoo's directors, Amon and Heidger, came from. The animals they donated included three species of monkeys, three peacocks, four poison toads, a de-scented skunk, and a macaw. The zoo was trying to get animals suitable for a hands-on, interactive environment for children. The newly refurbished flight cage (with a dirt floor instead of concrete) was filled with different species of birds, featuring two flamingos. (Above, courtesy of Richard Hammond; below, courtesy of Montgomery Zoo.)

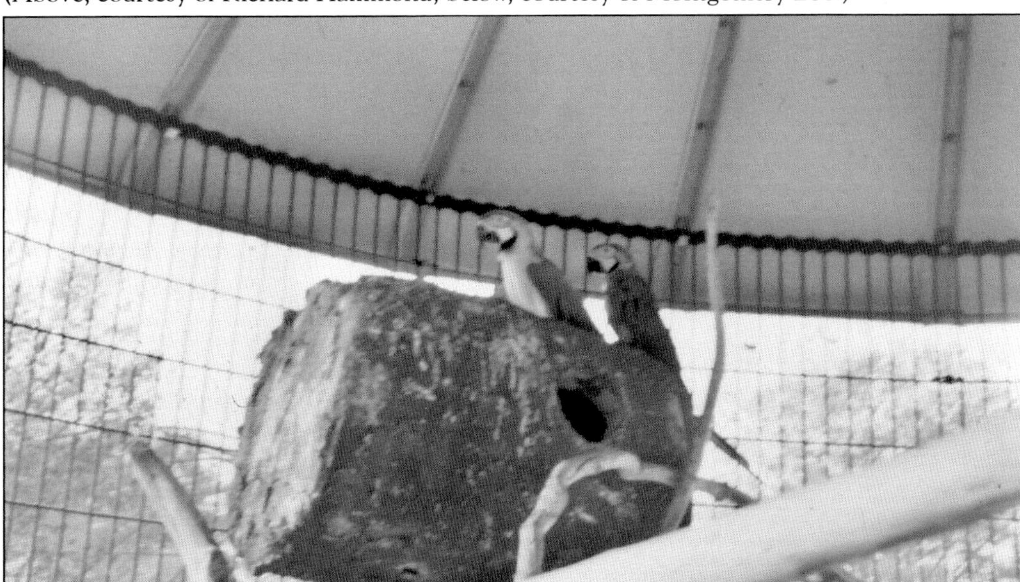

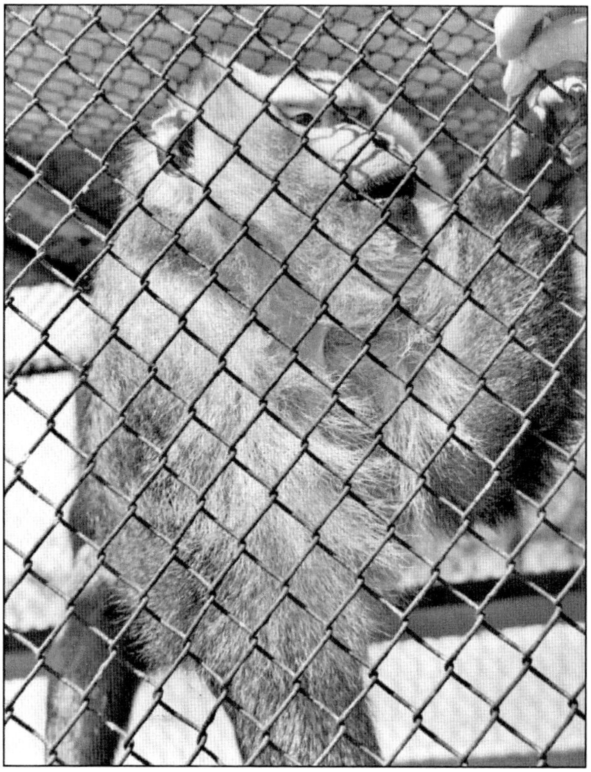

As visitors walked through the zoo, they could see the animals displayed in fully enclosed, round, wire cages (known as Baylen cages) or open-air paddocks. On the more appealing side, the Montgomery Children's Zoo had two species of macaques—a stumptail macaque and a pigtail macaque. They were displayed at the end of the zoo along with a bobcat, a raccoon and a greater crested porcupine. (Courtesy of Richard Hammond.)

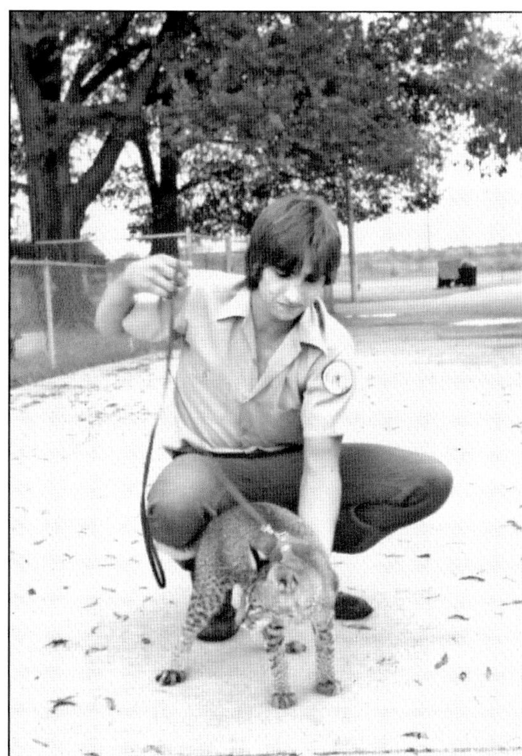

In the small 6-acre facility, the composition could only be classified as a menagerie, with 84 species acquired haphazardly. There was no organizing element that could be identified in the design of the facility and no education program. Animal food was scrounged from local grocery stores and other businesses. (Courtesy of Montgomery Zoo.)

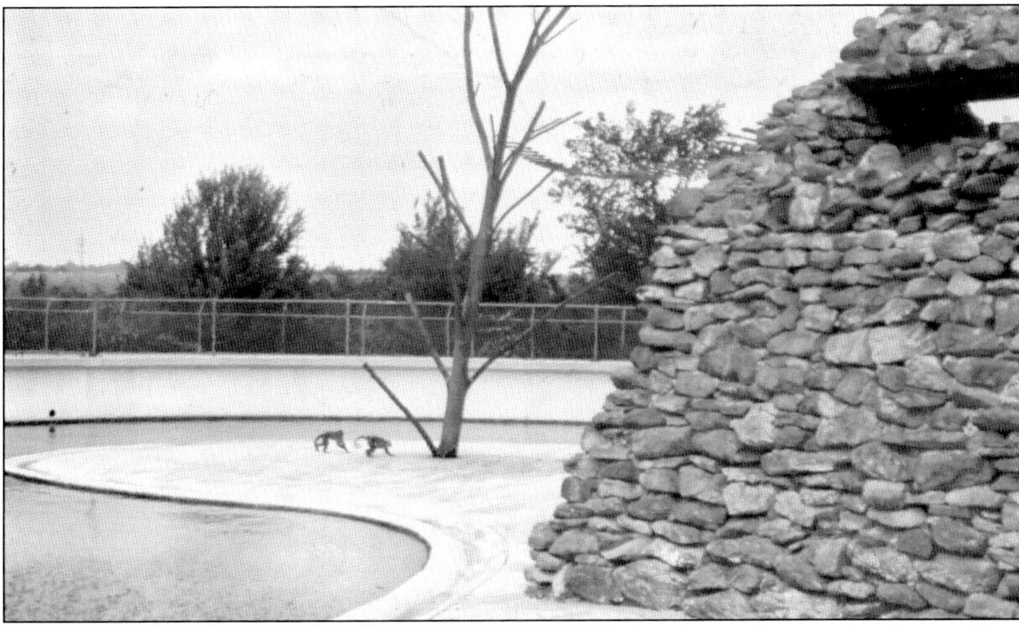

Monkey Island occupied much of the zoo's center and featured capuchins and spider monkeys. These highly energetic monkeys, whose antics on their two-bedroom heated house, and on the trees and bridges, would often detract visitors' attention from the alligator turtle in the moat below. The most well-known monkey on Monday Island was Red, a male capuchin. Originally the pet of a private owner, Red had his front teeth removed so that he would be unable to bite. However, the owner found he could no longer raise Red and donated him to the Montgomery Children's Zoo. Red was put on Monkey Island to live with the other monkeys. With no front teeth, however, he found it difficult to defend himself against the other monkeys. Red tried leaving the island by swimming across the moat. Eventually he was moved off the island to where he would be safe. (Courtesy of Montgomery Zoo.)

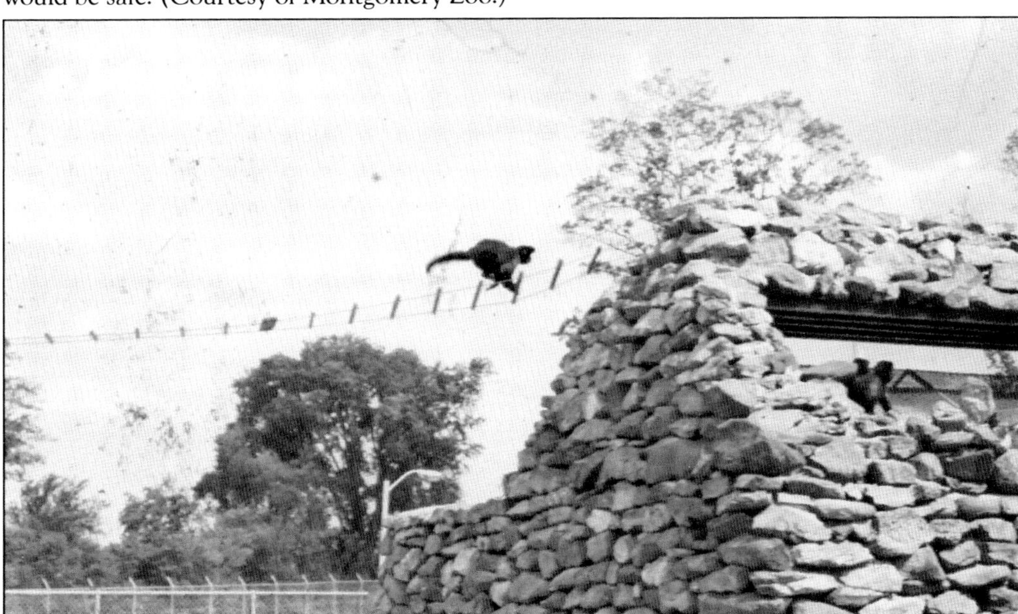

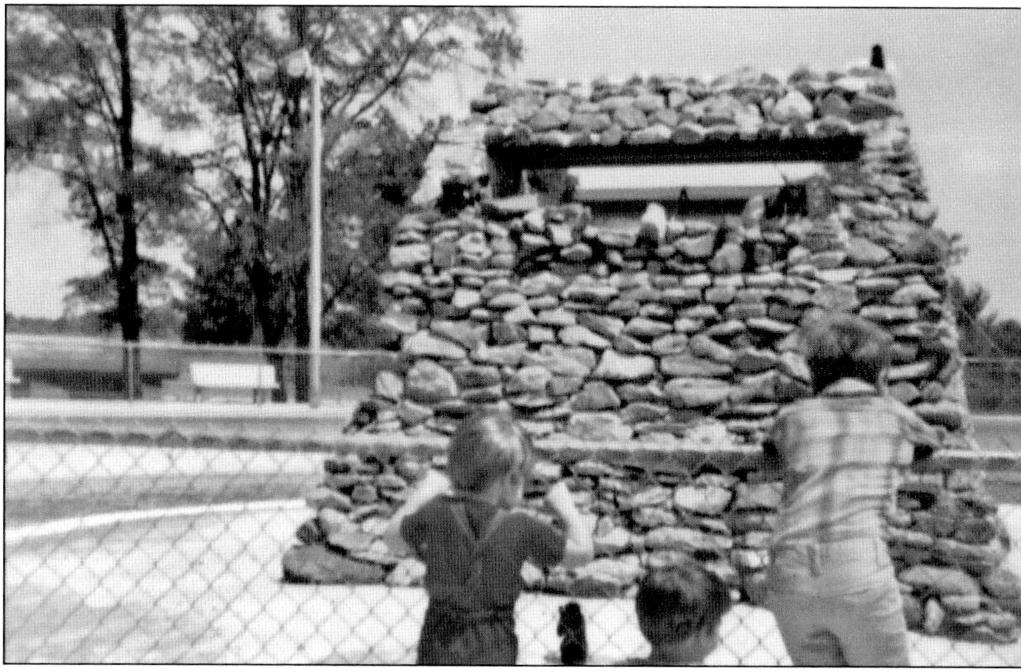

When Oak Park closed, all of the animals were either sold to private owners or donated to other zoos. Monkeys could be purchased at that time for only $35. When the Montgomery Children's Zoo opened in 1972, owners who felt like they could no longer care for any wild pets would bring them to the zoo. Another notable primate of Monkey Island was a female capuchin that had been purchased from the original Oak Park Monkey Island group. (Above, courtesy of Betty Breslin; below, courtesy of Montgomery Zoo.)

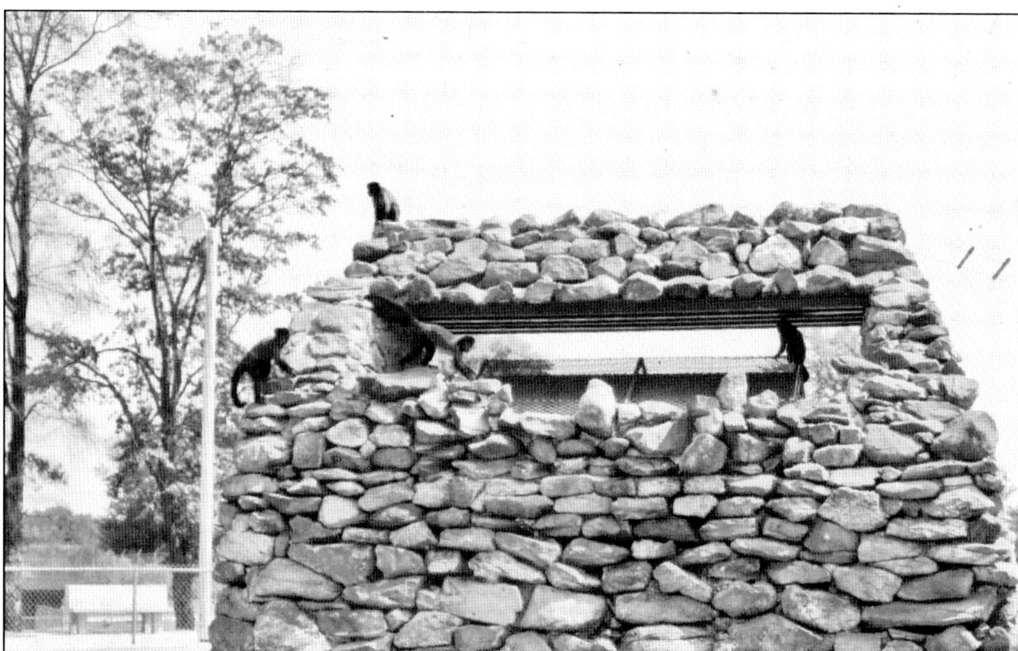

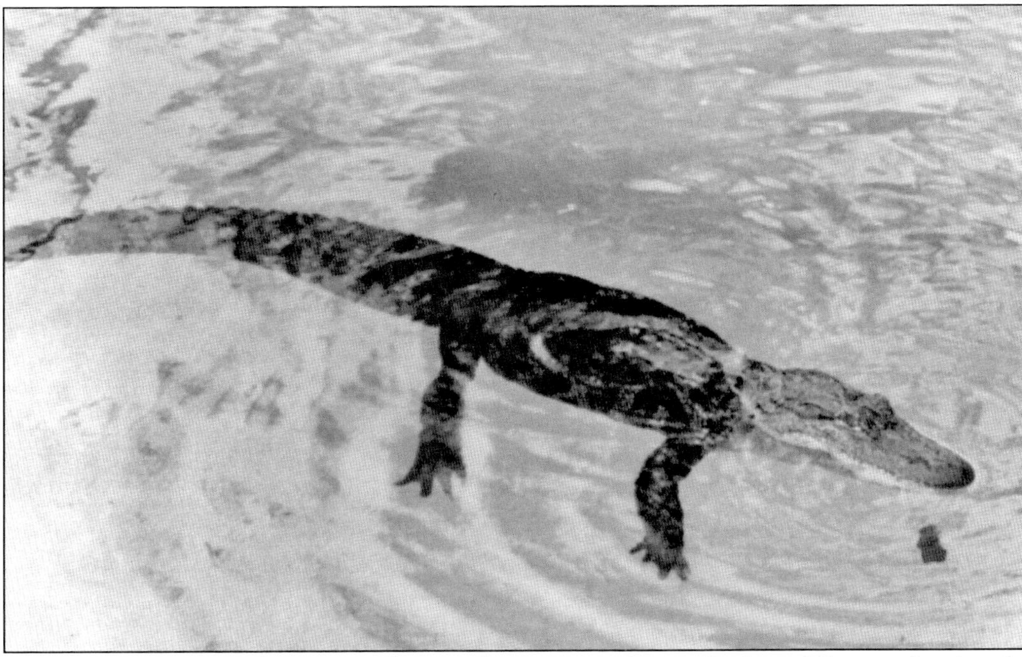

Monkey Island provided visitors with more than just monkeys to look at. Alligators lived in the moat surrounding Monkey Island. There they sunned on the concrete and dreamed that a monkey would fall into snapping reach. The zoo named its largest alligator Number One. He lived in the moat with smaller companions, alligator turtles. (Courtesy of Montgomery Zoo.)

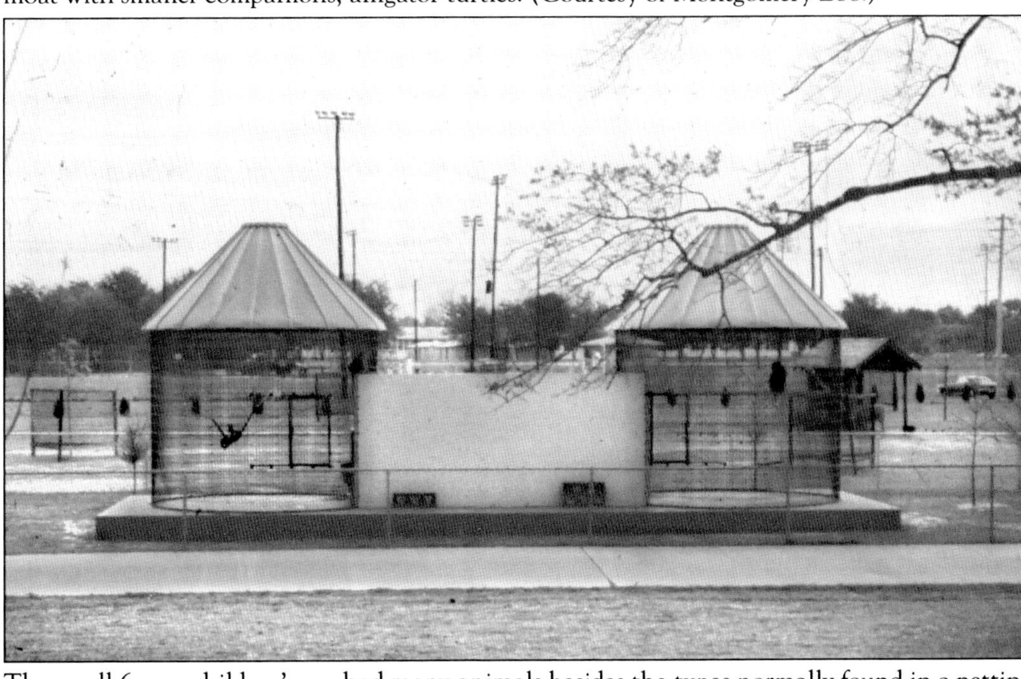

The small 6-acre children's zoo had many animals besides the types normally found in a petting zoo. A short hop from the walk-through flight cage were two Baylen cages. Inside these cages, children would discover two spider monkeys known to the keepers as Perry and George. (Courtesy of Montgomery Zoo.)

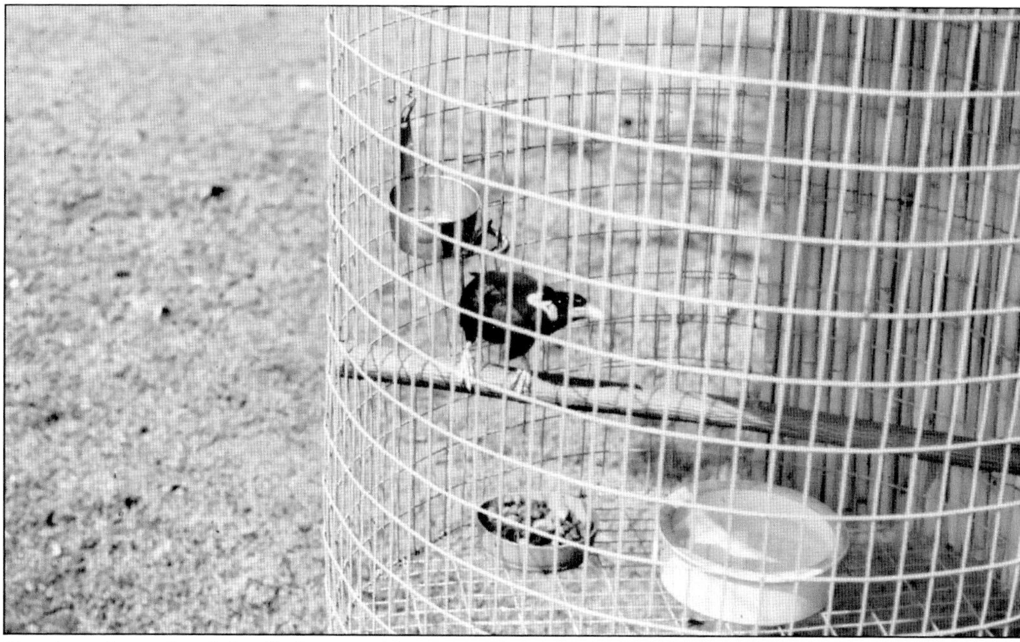

On the other side of the flight cage, the petting area emerged. Housed here within paddock-type areas were the llamas and rheas, both South American natives. Here children could lure the animals close with food that they bought at the zoo, but they had to be careful not to upset the llamas or they would get spit on. Feeding rheas can be a tricky task as well, as the children soon learned; even though there are no teeth in that beak, if they snap one's fingers while taking the food, it can be a painful experience. Along the path, children would encounter the fascinating mynah birds, whose mimicking abilities could keep one in conversation for hours. (Above, courtesy of Betty Breslin; below, courtesy of Montgomery Zoo.)

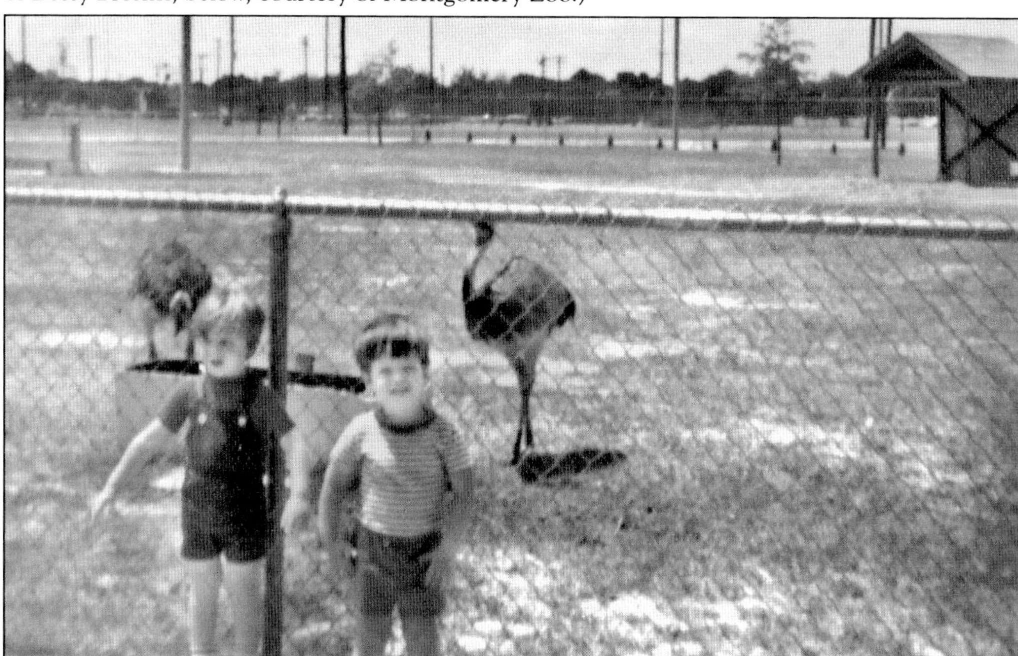

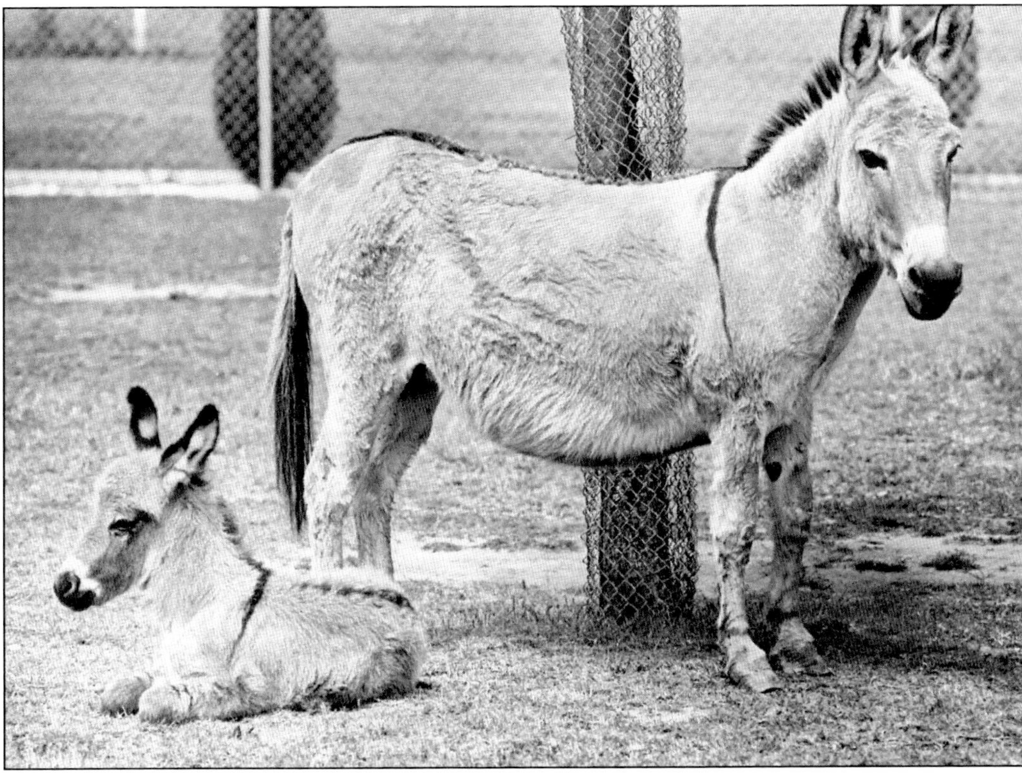

Sweet Thing, a small donkey, was one of the first animals in the petting zoo. She had been the unofficial mascot for nearly a year at the Air University's Air Command and Staff College at Maxwell Air Force Base. She found a permanent home at the Montgomery Children's Zoo. (Courtesy of Jerry Smith.)

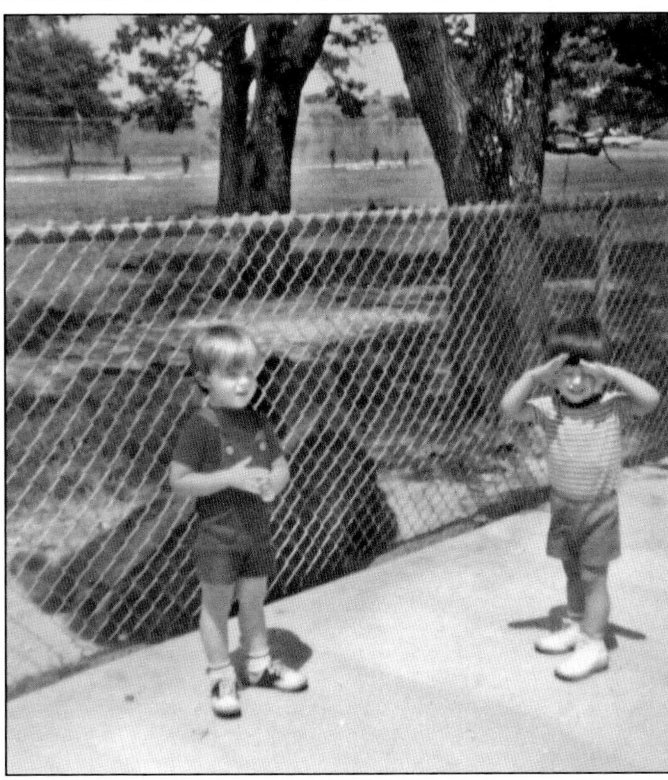

Bashful, Boulder, and Bramble were the three North American buffalo (also commonly known as bison) living at the Montgomery Children's Zoo. Children loved to watch these majestic symbols of the West. Lamar (left) and Andrew (grandsons of a Montgomery Children's Zoo volunteer, Betty Breslin) posed for a photograph in front of the buffalo enclosure. (Courtesy of Betty Breslin.)

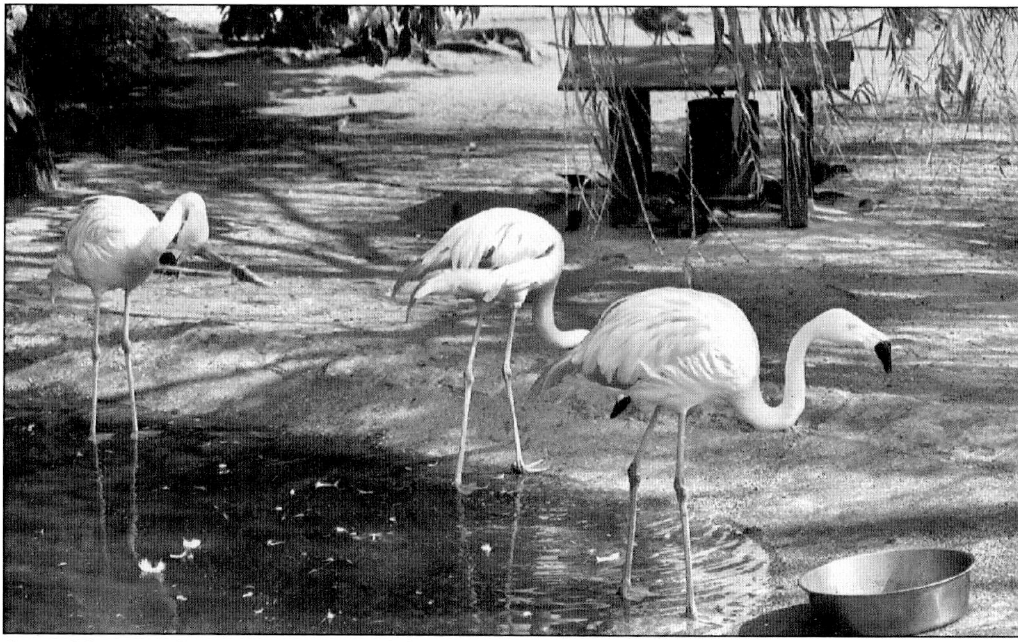

One of the most fascinating aspects of the Montgomery Children's Zoo was the huge aviary, bustling with some 25 species of birds nestling and going about their business in spite of the proximity of spectators who could walk through the aviary over an arched bridge. Providing special interest were large flamingoes. (Courtesy of Marsha T. Lautenbach.)

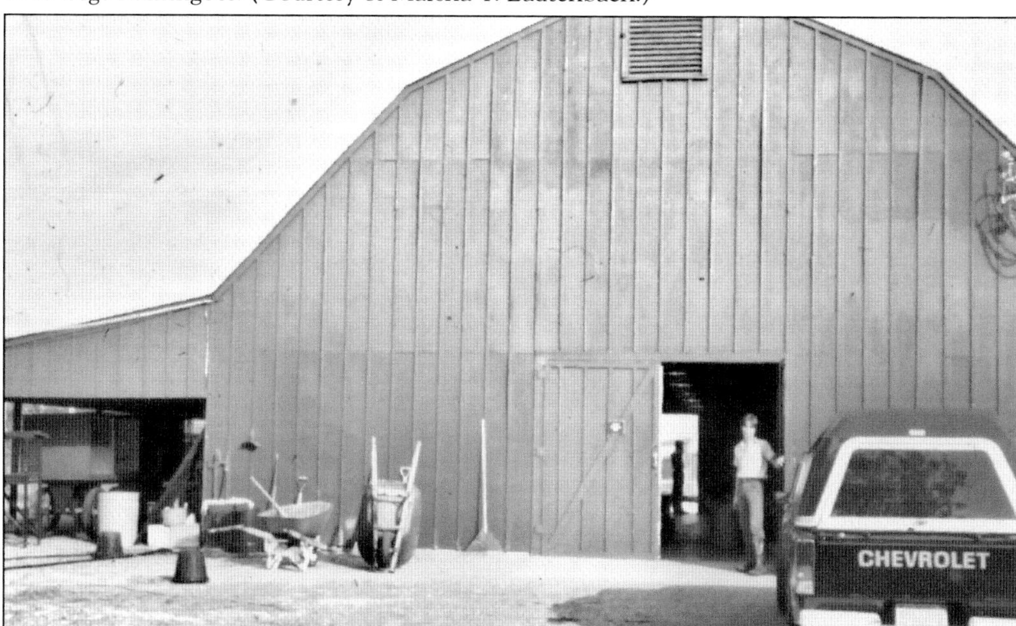

Billy, the zoo's first chimpanzee, was housed in the red barn because the Montgomery Children's Zoo had no facilities to hold him. He would walk around on a chained leash with John Heidger. Billy was soon traded because of the lack of appropriate housing. The zoo was to receive a bear in exchange for Billy; however, the zoo was unable to accept it because of facility problems again. (Courtesy of Montgomery Zoo.)

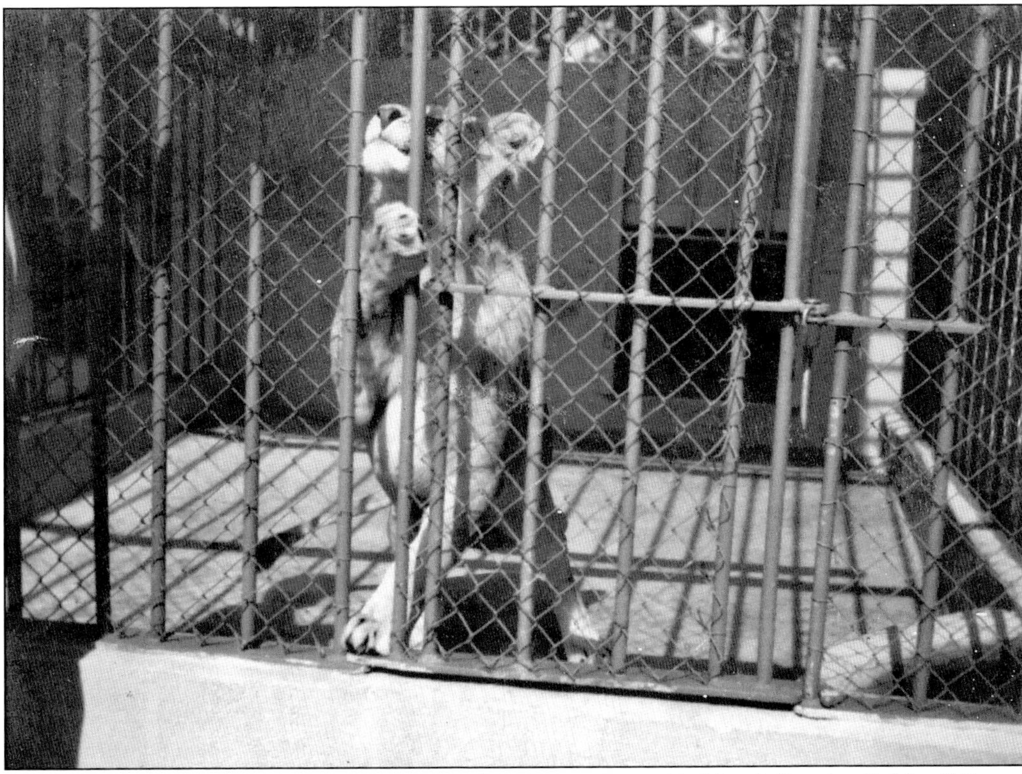

In 1971, Myrobi, a male lion, was given to the Montgomery Children's Zoo by Caldwell Zoo in Tyler, Texas. Myrobi was only six months old. The zoo introduced Myrobi to Big Mama, an older female lion born in 1959 that came from an Atlanta zoo. (Courtesy of Montgomery Zoo.)

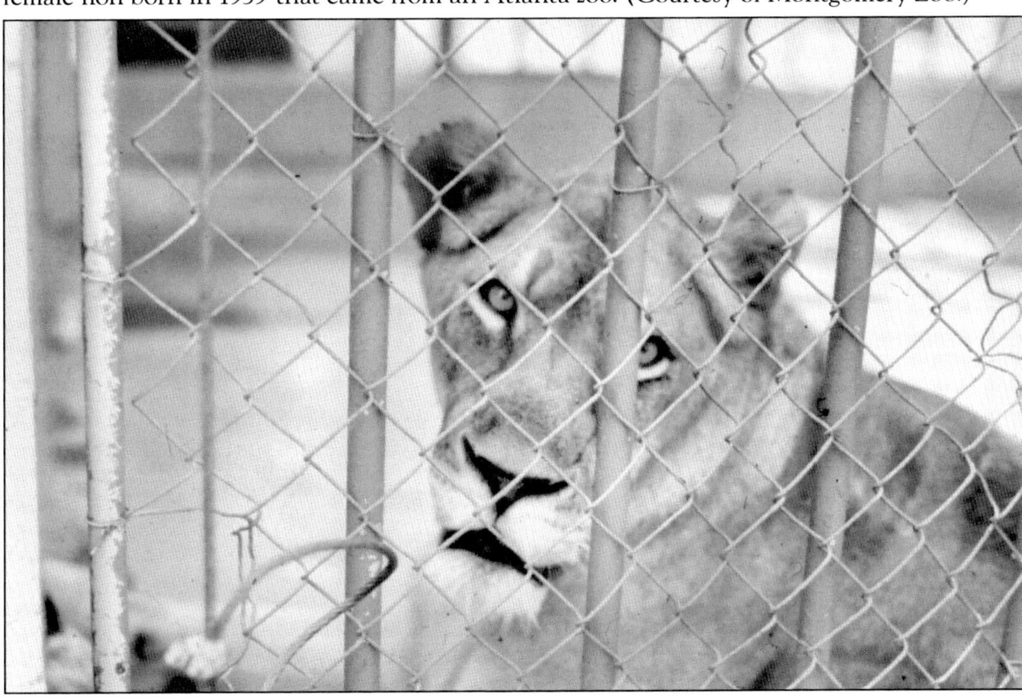

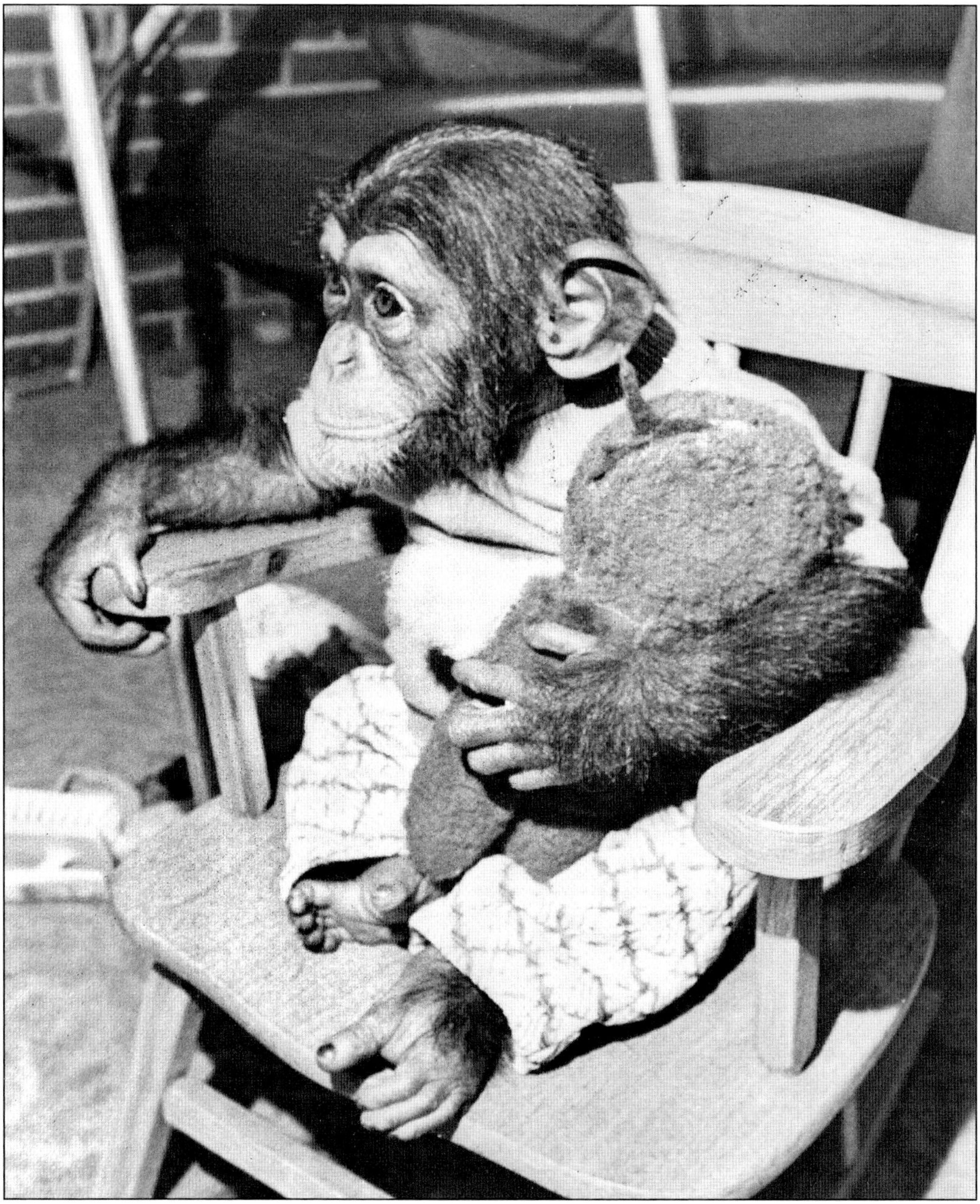

In an effort to expand the types of animals available for the public to see, the zoo received its first baby chimpanzee in 1972. Benji (although his official zoo name was Binjo), a male chimp, was picked up from the Miami airport by zoo director John Heidger in August that year. The recent addition to the zoo came from "the wilds of Africa." Being just a baby—Benji was less than a year old—every move of his was watched. Since he was too young to stay at the zoo alone, he was carried home each evening for safe keeping to the house of a zoo staffer. The zoo did not have any other chimps for Benji to live with; therefore he was hand-raised. Benji's personality was shaped by this constant attention. (Courtesy of Fran Bergquist.)

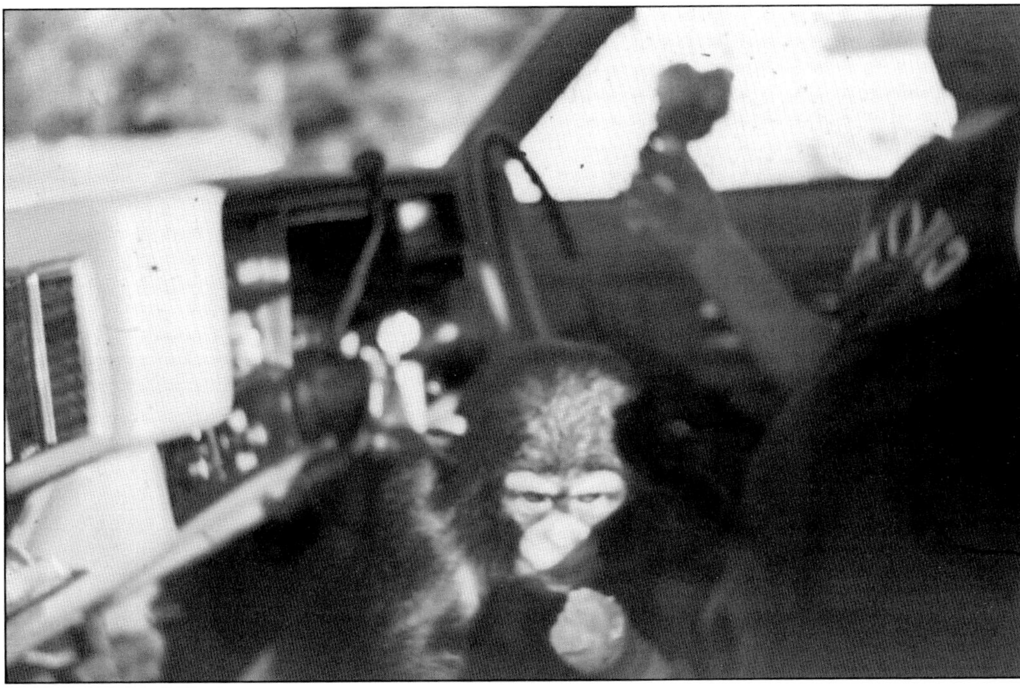

Benji was an unusual "child" that lived with Joe and Laura Strickland of Montgomery. He was too young to live at the zoo, so Laura Strickland, who was the zoo director's assistant, provided Benji a home. Benji quickly became one of the family. He would partake in everything the Strickland children would do. He would eat ice cream, ride back and forth from the zoo in a car, be carried around like a baby, take bubble baths, play with the family's cats and dogs, brush his teeth before bed, and even wear pajamas to sleep. (Courtesy of Fran Bergquist.)

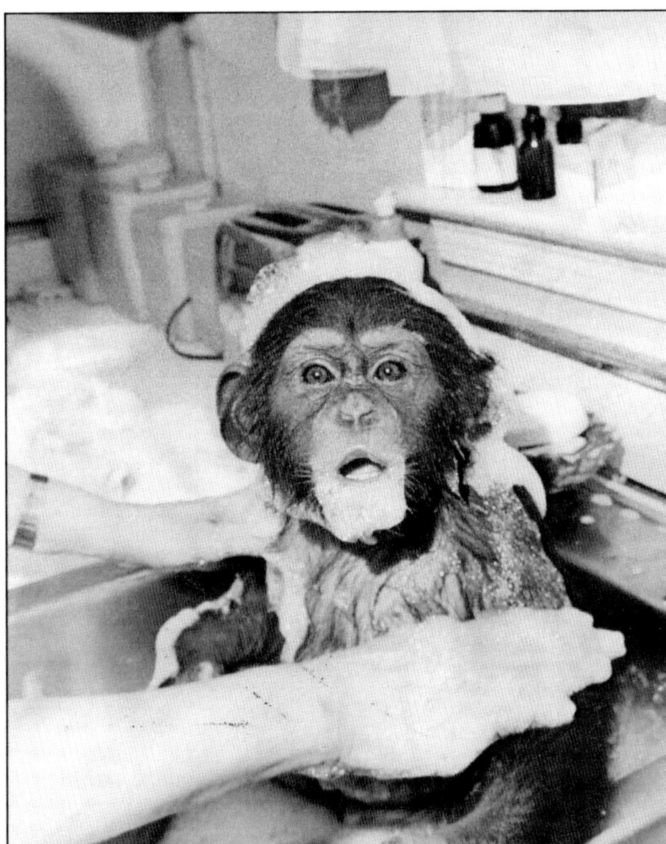

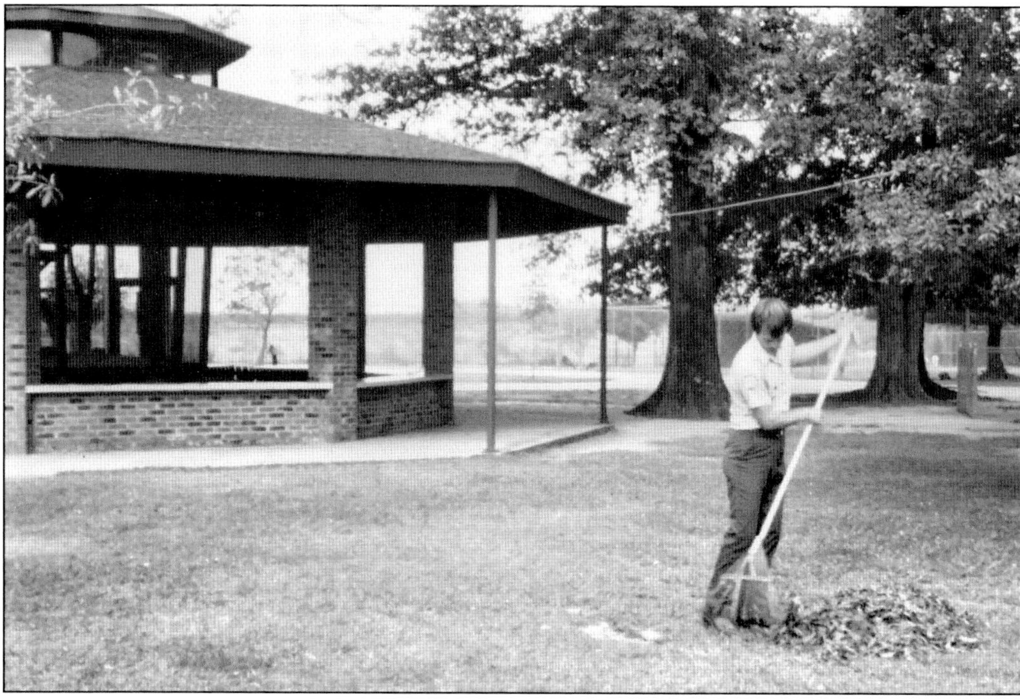

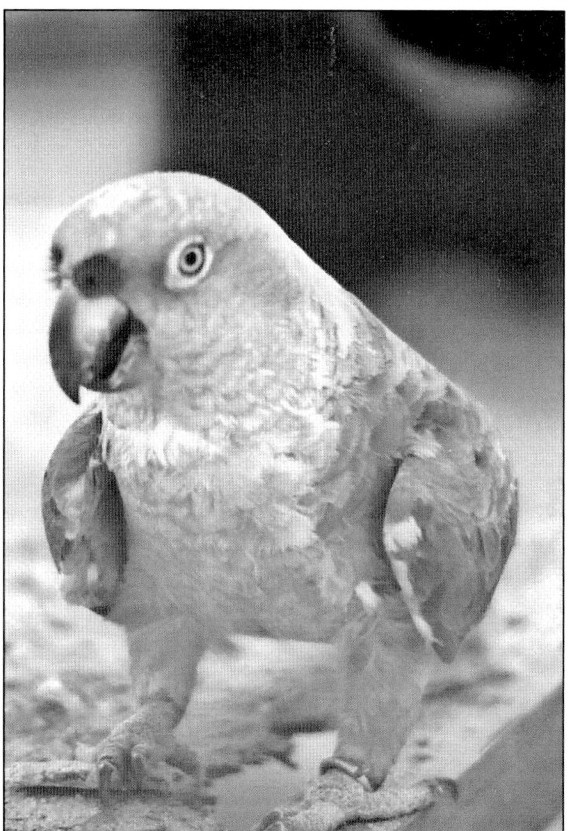

To keep the zoo an exciting place to visit, plans for new exhibits started shortly after the children's zoo opened. One of the first would be an innovative educational shelter where guests could interact with animals and educators at the same time. The ground-breaking for the sheltered petting area was in November 1972, though it would not be completed for a few years. Plans for the circular structure included a center that would house tropical birds surrounded by enclosed animal pens. Here children would be able to have monitored encounters with these demonstration animals. To make these plans come to be, additional funds were raised by the Capitol Junior Woman's Club, who sponsored an 8-mile march from the State Capitol to Oak Park, the site of Montgomery's original zoo, and then to the new zoo. (Courtesy of Montgomery Zoo.)

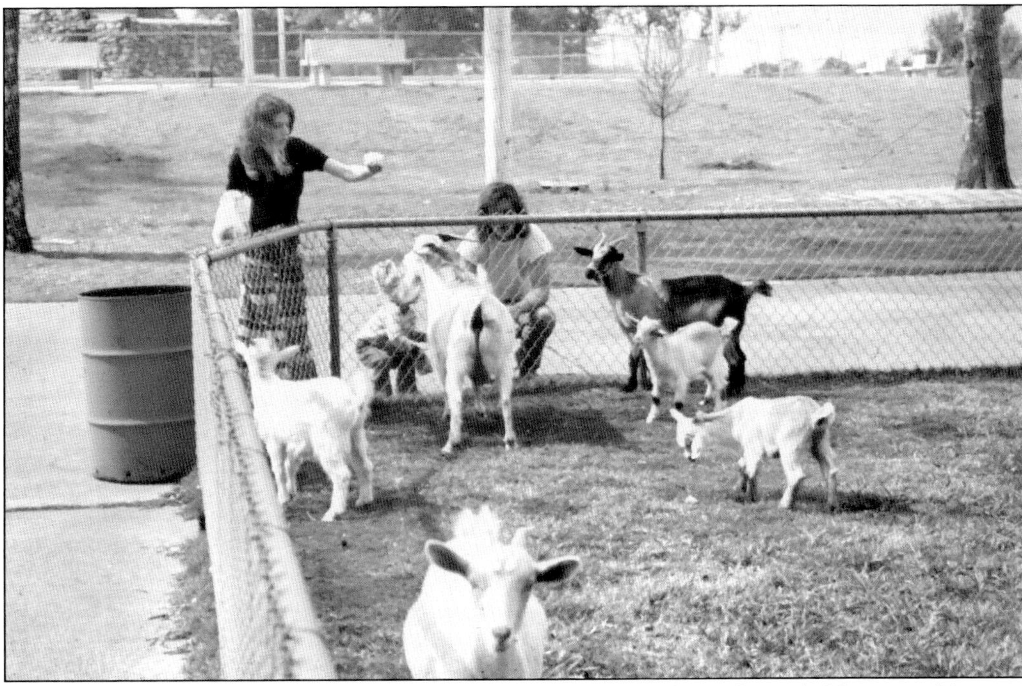

At the heart of any children's zoo is the petting zoo area, the place where children are able to get up close and personal with the animals. The petting zoo at the Montgomery Children Zoo hosted a variety of larger domesticated animals such as sheep, goats, and llamas. Several species of deer, such as exotic axis and fallow deer, were exhibited among the local whitetail deer. Children could feed the deer through the fence and learn about not only their local wildlife, but also see the similarities and differences between the wildlife in their backyard and other childrens' backyards around the world. (Courtesy of Montgomery Zoo.)

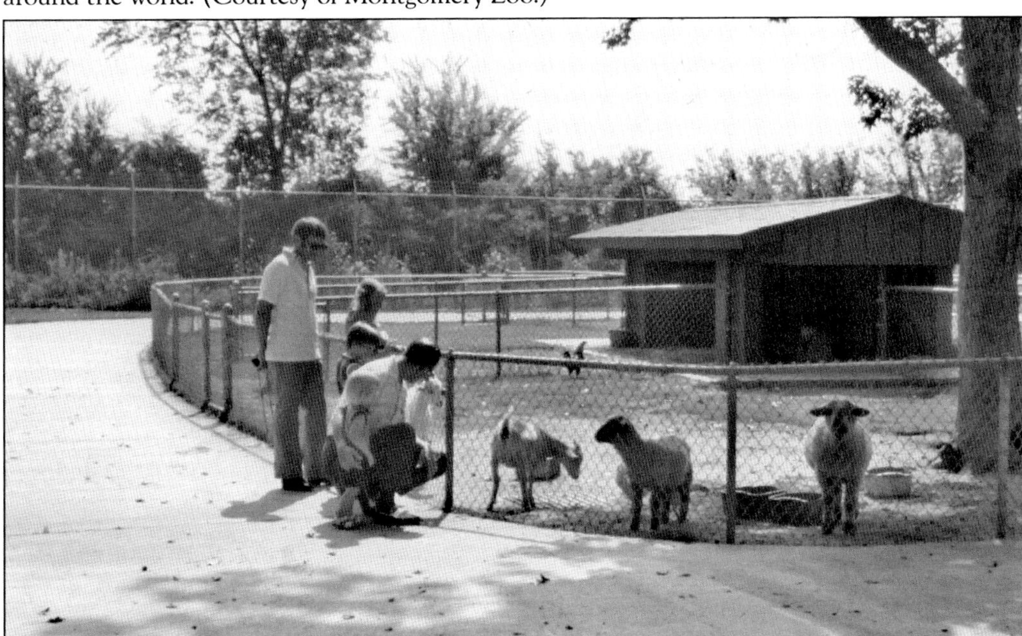

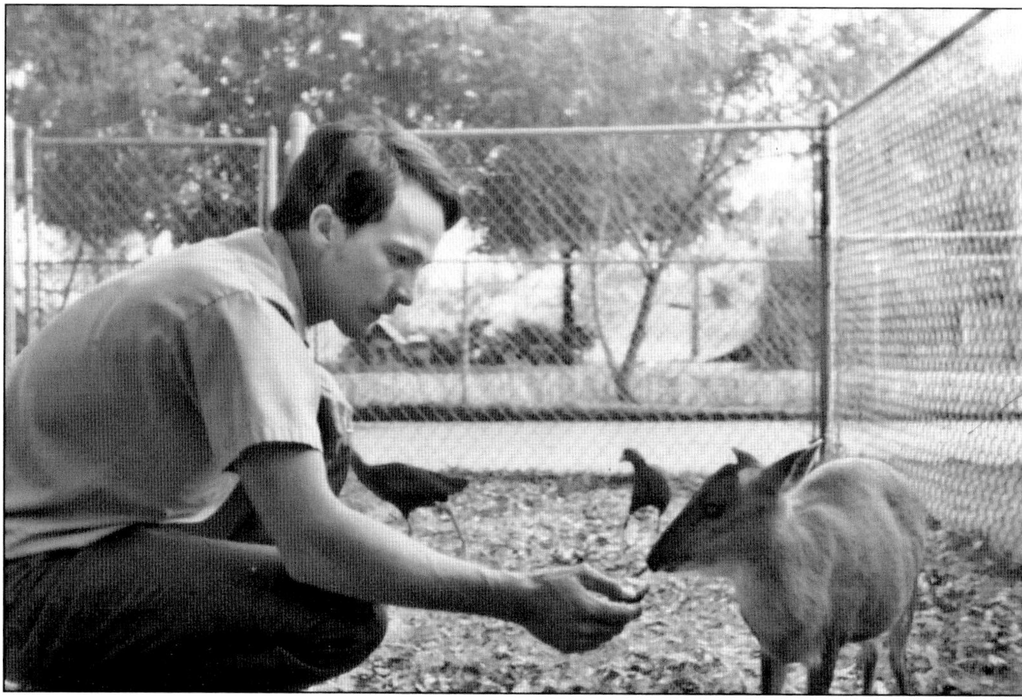

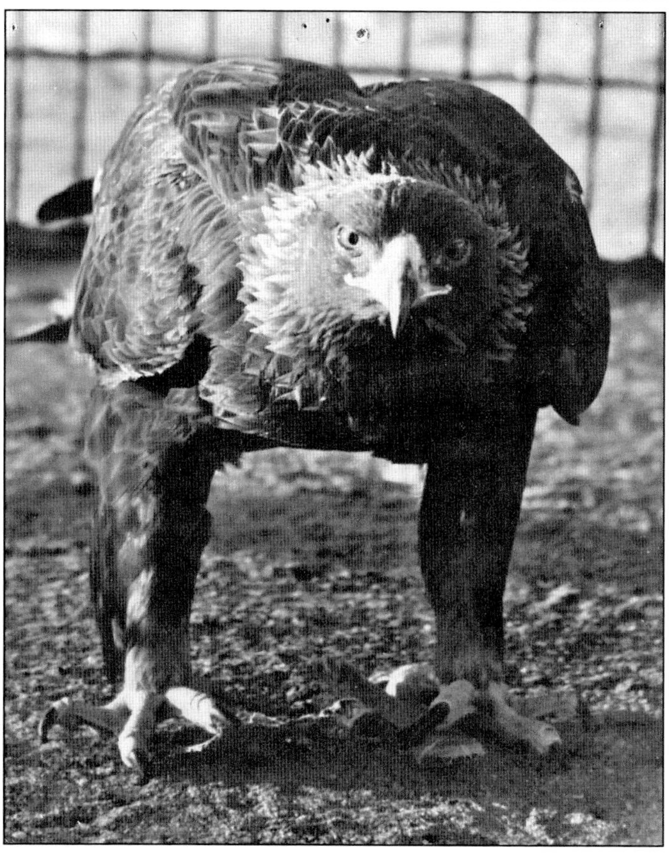

The Montgomery Children's Zoo employed a variety of people that liked to work with animals. Paul Foster, Jim Rogers, and Mike McGough (pictured above with a muntjack) were zookeepers. Lynelle and Mrs. Taylor worked the concession stand. Secretary Laura Strickland maintained the office and kept zoo director John Heidger organized. (Courtesy of Montgomery Zoo.)

The golden eagle, named for the goldenish feathers on its back, was an educational animal at the zoo. Golden eagles were added to the Bald Eagle Protection Act in 1963, meaning that no one could take one from the wild. However, if a local bird was not able to survive in the wild, it was given to the zoo to become an ambassador for its wild counterparts. (Courtesy of Robert B. Wallace.)

Some 20 well-wishers turned out on December 5, 1972, to celebrate the first birthday of West Park Avenue's most unusual resident, Benji the chimpanzee. Benji celebrated a more informal party at Laura Strickland's home with his closest friends. Along with two stuffed animals, Benji received cakes, cookies, and other toys, including an inflatable Santa Claus, which came to a swift end at the teeth of the chimp. (Courtesy of Fran Bergquist.)

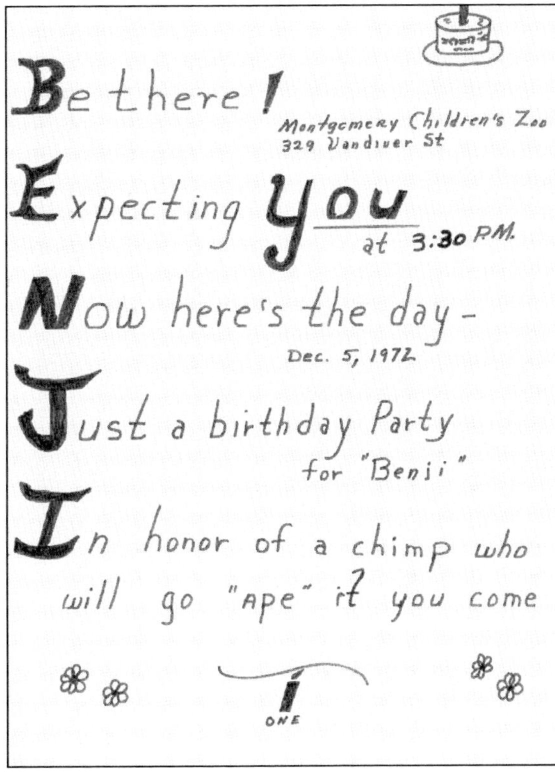

Be there!
Montgomery Children's Zoo
329 Vandiver St
Expecting you at 3:30 P.M.
Now here's the day - Dec. 5, 1972
Just a birthday Party for "Benji"
In honor of a chimp who will go "ape" if you come!
ONE

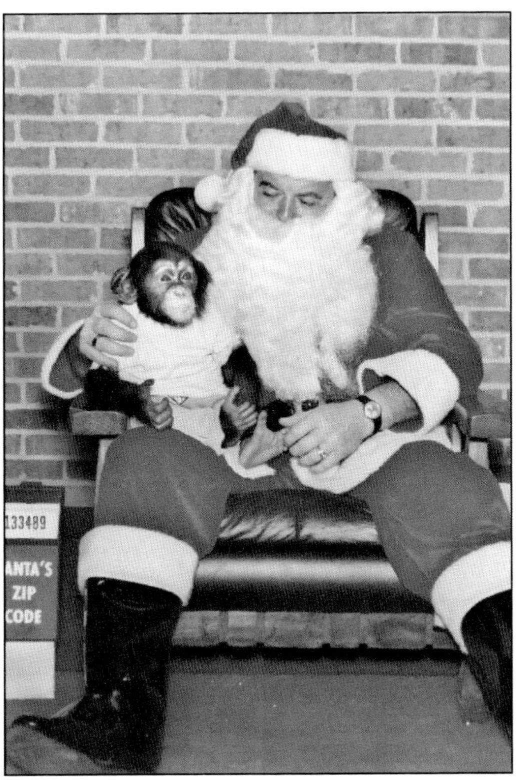

Benji liked to do the same things that little kids liked to do. As part of a promotional event for the Montgomery Children's Zoo, Benji participated in all the Christmas traditions. He had a visit with Santa and sat on his lap like all the other children did. (Courtesy of Montgomery Zoo.)

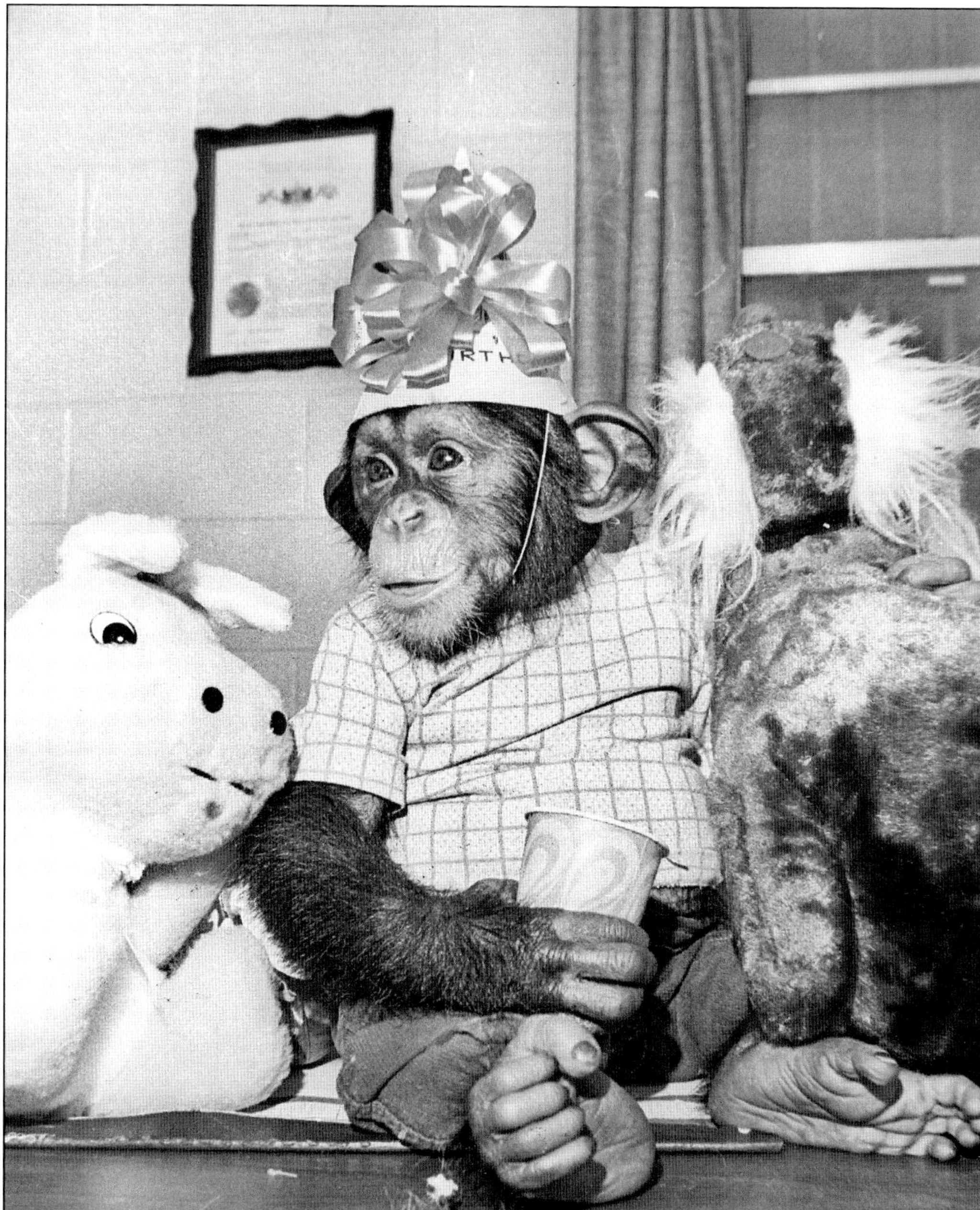

The Montgomery Children's Zoo also invited the general public to join in on the celebration of Benji's first birthday in December 1972. Following his more intimate party at his home, Benji was the spotlight for a gathering at the zoo. Celebrating in style, over 200 visitors brought presents for the unusual birthday boy and food for him to enjoy. To show his birthday spirit and join in on the festivities, Benji even wore a personally made birthday cap and partook in all the special treats brought by visitors. A special birthday cake topped with vegetables was prepared just for Benji by the zookeepers. Because of the popularity of the chimp display at the zoo, large birthday celebrations became a yearly event. (Courtesy of Montgomery Zoo.)

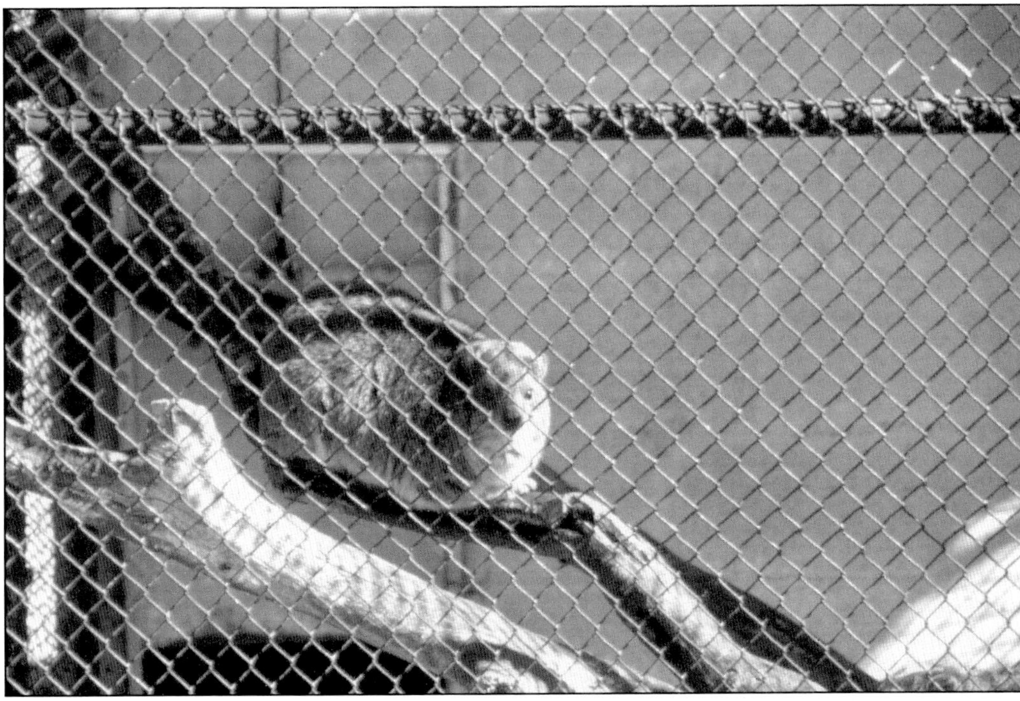

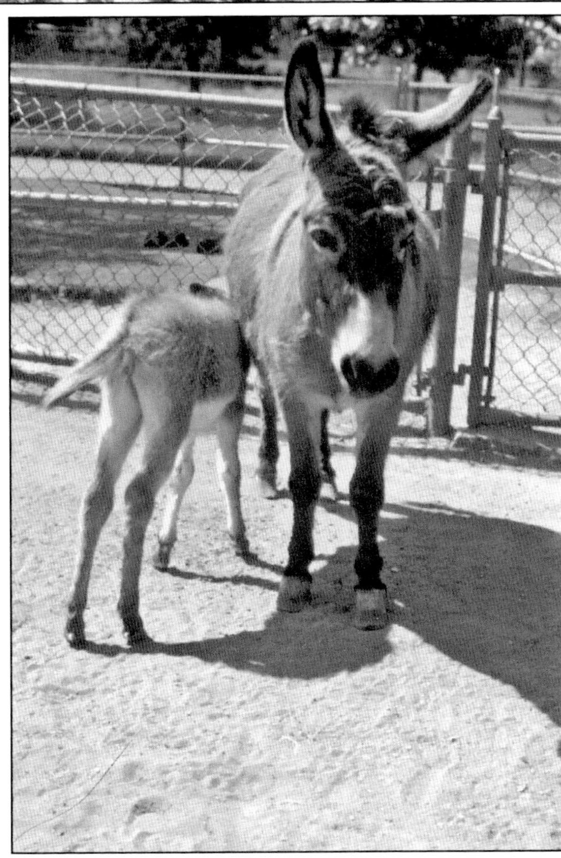

Most of the animals that were exhibited at the Montgomery Children's Zoo were displayed in pairs or in groups. This allowed for many of the newer animals to be either born or hatched at the zoo. The more notable births at that time included a squirrel monkey, hyrax, and Abyssisinian ass (commonly known as a donkey). Babies are always a must-see for visitors, especially children. Eventually one of the zoo's missions became maintaining breeding groups of animals, especially for endangered or threatened species. It is also a way for the zoo to increase or sustain its number of animals. (Courtesy of Montgomery Zoo.)

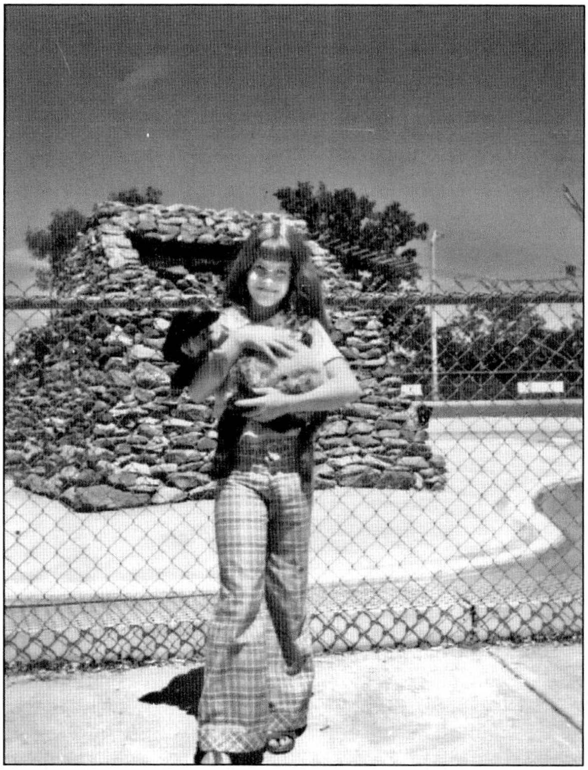

The first anniversary of the Montgomery Children's Zoo was celebrated on May 5, 1973. The festivities included a "Name the Baby Lamb" contest, art auction, animal safari, and artist party. Admission to the extravaganza was free for all visitors. (Courtesy of Montgomery Zoo.)

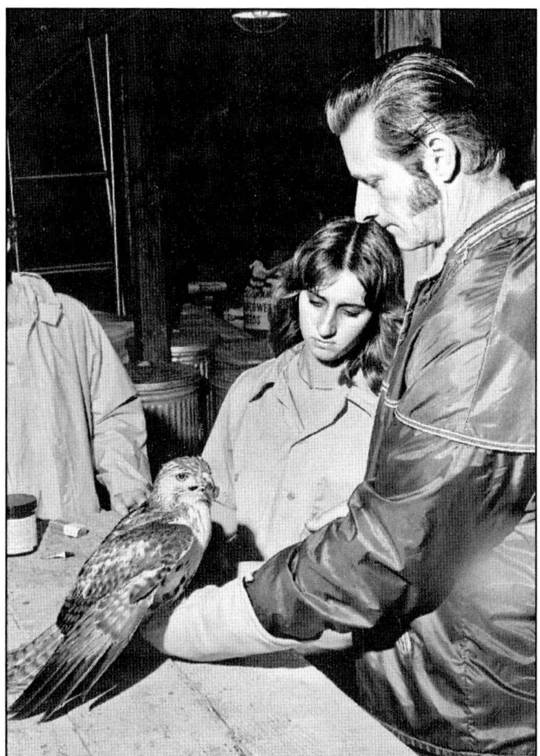

Due to health issues, John Heidger, the zoo's director, resigned his post in 1973. While the zoo searched for a new director, longtime keeper and now zoo curator Paul Foster (pictured on the right) became interim director again. The zoo began its third search for a new zoo director within three years. (Courtesy of Charlie Steed.)

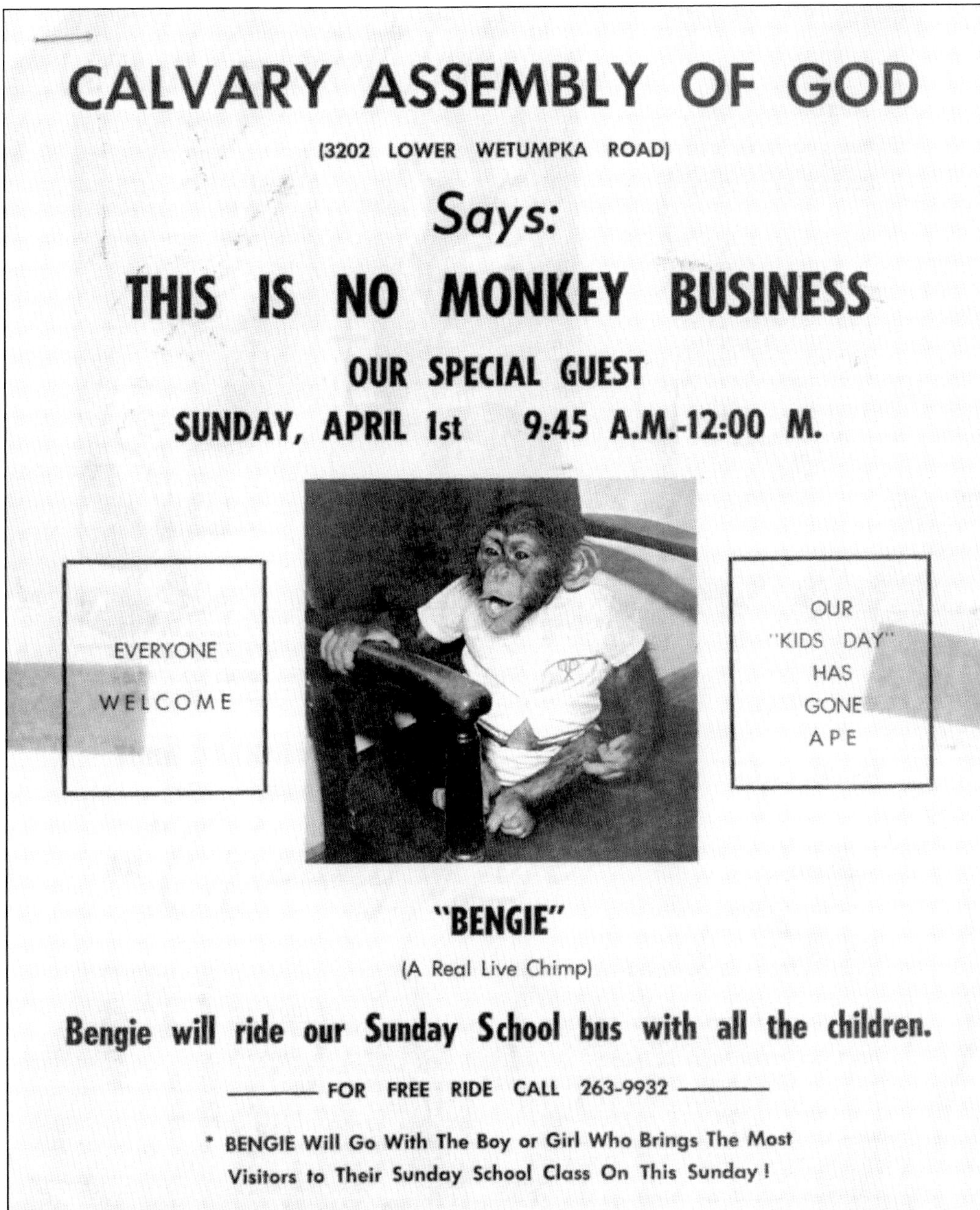

Just like his visit with Santa, Benji, whose name is misspelled in the above advertisement, continued to participate in normal childhood activities. Calvary Assembly of God's "Kids Day" went ape on Sunday, April 1, as Benji visited. His visit was entitled "This is No Monkey Business." He rode the Sunday school bus with all the children to church that Sunday. To encourage participation in the church, the boy or girl who brought the most visitors to Sunday school class that day was able to spend the rest of the morning with Benji. The children were excited to be able to play with a chimp. The zoo also used this opportunity to increase awareness about chimpanzees and other animals that the zoo had to offer. This was one of the first educational visits for an animal outside of the zoo. (Courtesy of Calvary Assembly of God.)

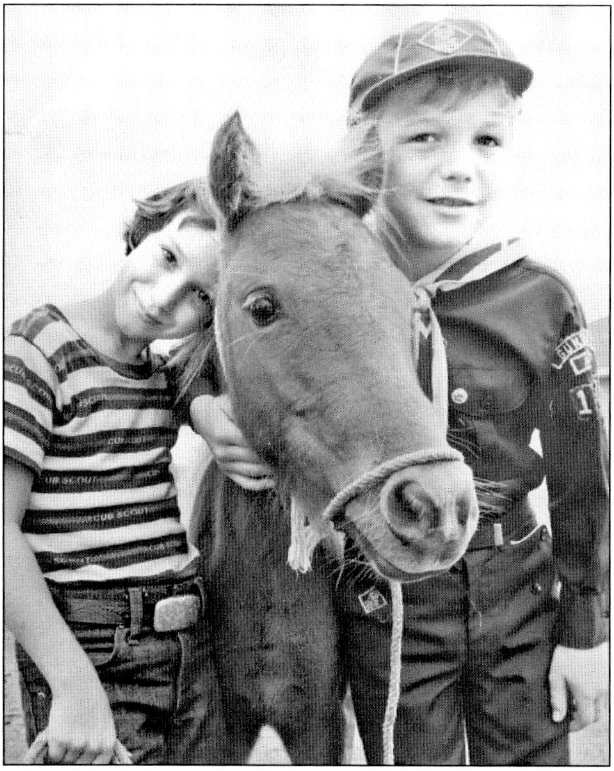

In September 1973, the Montgomery Children's Zoo sponsored "Wild Kingdom Day." Five hundred city Cub Scouts came to the zoo and participated in the event. A raffle prize for the kids included a pony donated by Sonny Milstead. Ten-year-old James Adams's name was drawn as the winner of a Shetland pony. (Courtesy of Montgomery Zoo.)

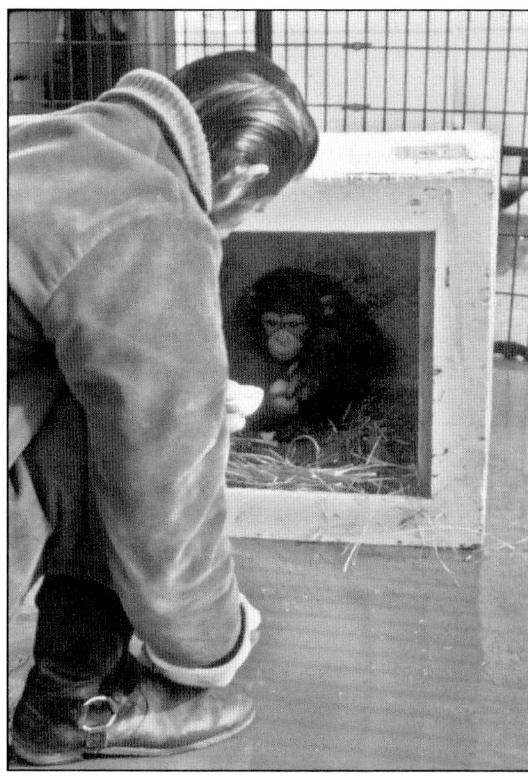

Benji got a Christmas gift of a new girlfriend in December 1973. Little Liza Jane arrived at the Montgomery Children's Zoo and was introduced to her new playmate. Liza Jane was brought from Africa to a nursery at a drive-through animal park. The zoo had been looking for a companion for Benji for some time. (Courtesy of Montgomery Zoo.)

C. B. "Bird Dog" Rogers was hired as director of the Montgomery Children's Zoo in April 1974. His future plans included completing a gator pool, waterfowl display area, and a permanent display area for birds. Before coming to the zoo, he was a former education specialist for the Oklahoma State Department of Wildlife Conservation. (Courtesy of Montgomery Zoo.)

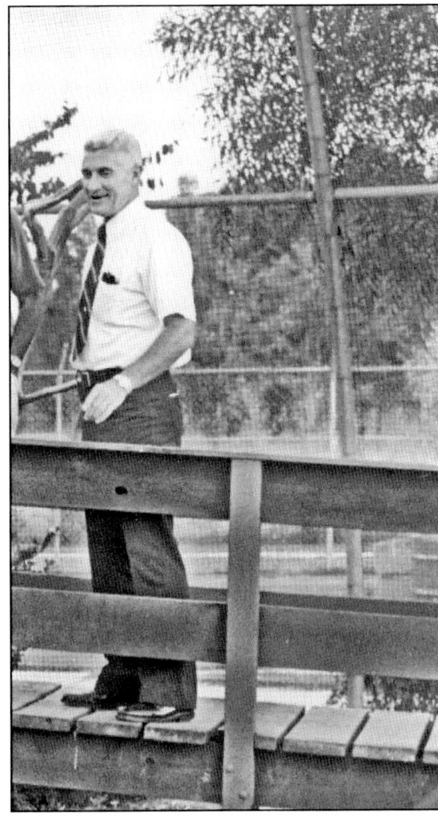

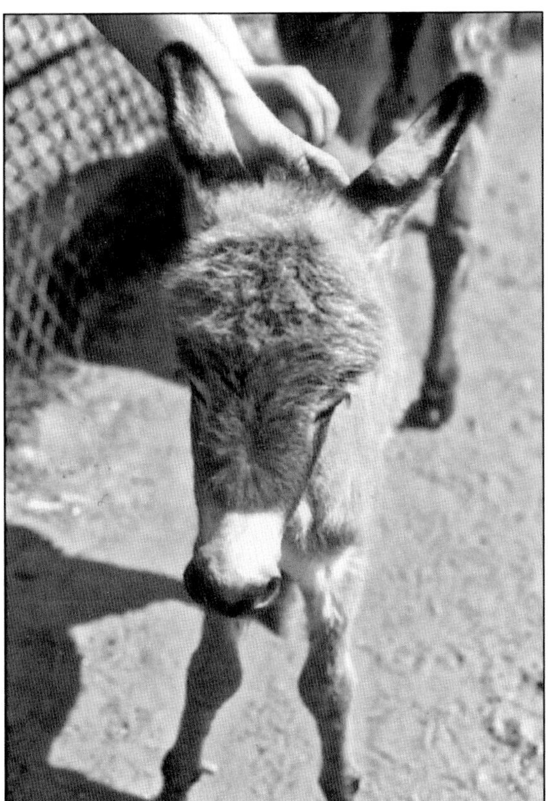

The Montgomery Children's Zoo celebrated its second birthday on May 4, 1974. A special contest was held to name the new baby donkey. Admission was free. Special activities occurred throughout to the day, including a painting contest of the animals in the zoo. (Courtesy of Montgomery Zoo.)

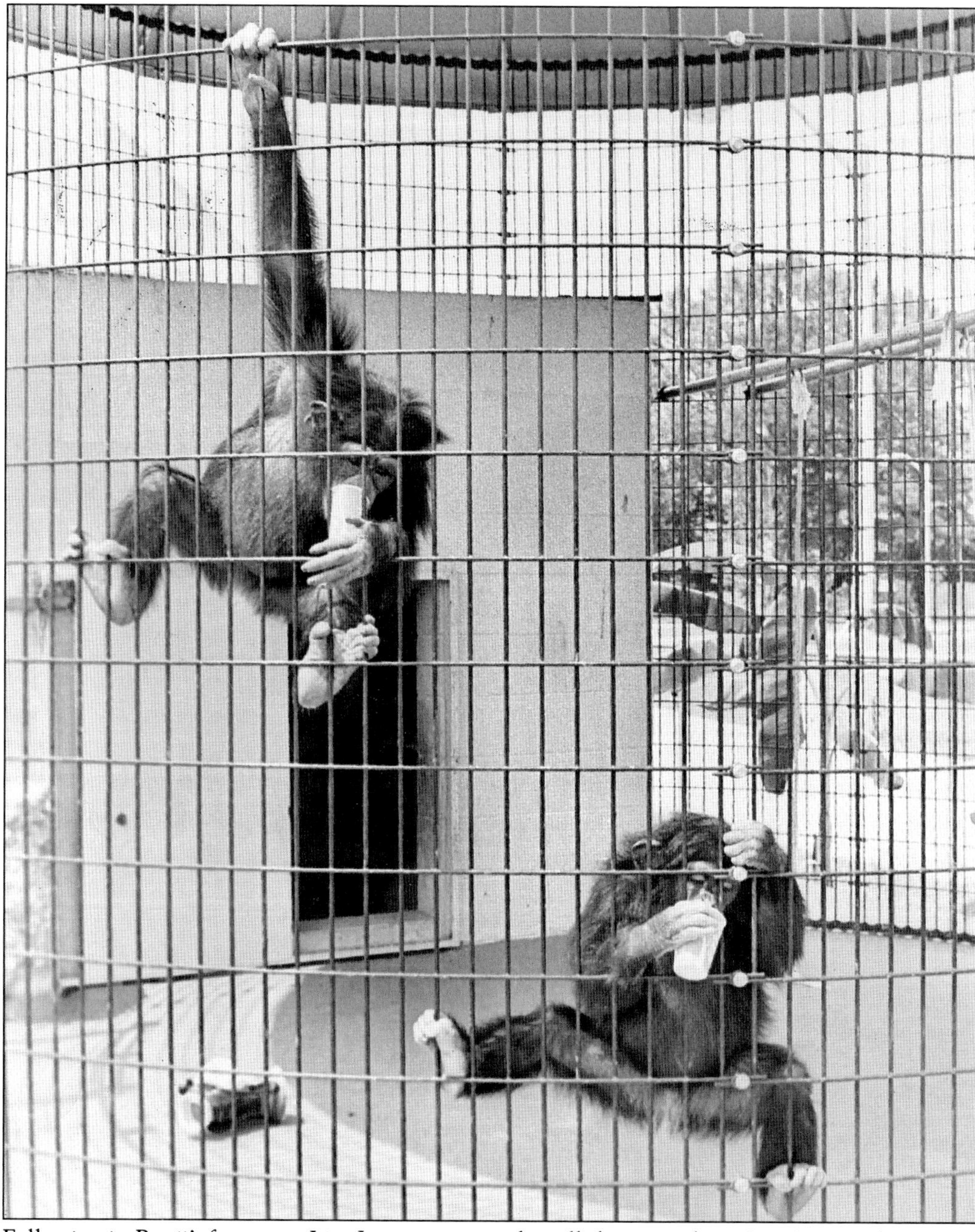

Following in Benji's footsteps, Liza Jane was treated to all the specialties. Liza Jane and Benji got an iced-cold soft drink served in a cup every afternoon during the summer at the Montgomery Children's Zoo. The visitors loved watching the chimps drink their sodas, and they would be aped by some of the smaller visitors to the chimp cage. Liza Jane and Benji became immediate friends and enjoyed each other's company. They would not only entertain each other but also the public with their childish antics. Chimpanzees held a special place at the zoo and were treated like royalty. The zoo wished to expand their small group but lacked the necessary facilities. Even these chimps would one day become too big for their current enclosure. (Courtesy of Roy Garvin.)

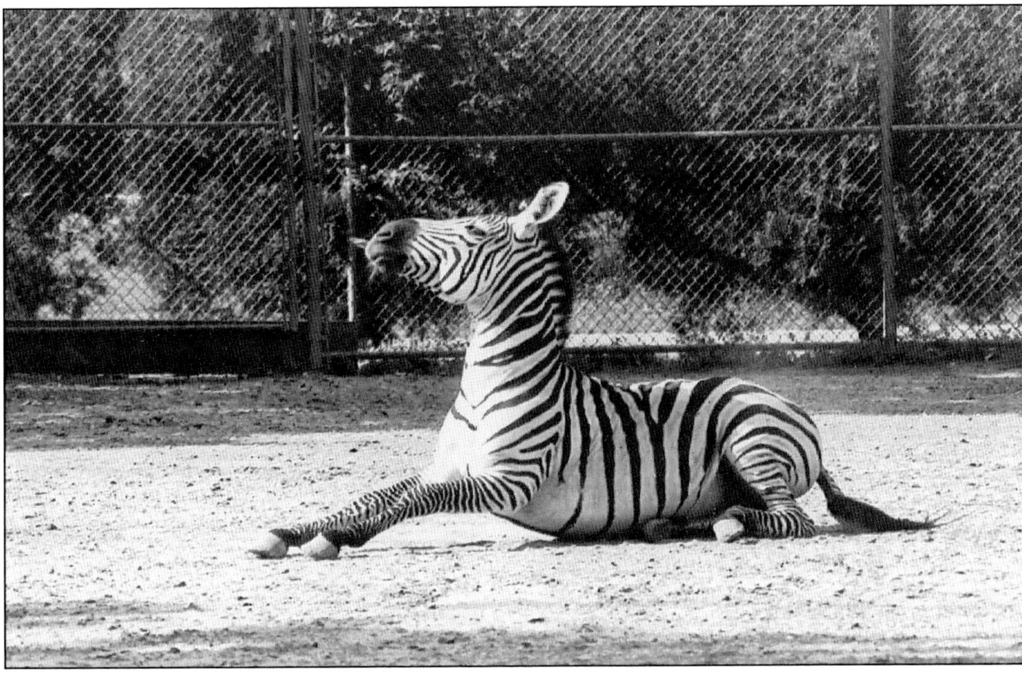

To make the petting area look wilder, less domestic, more exotic, yet tractable animals were wanted for addition to the zoo. After a search for available animals, two African zebras (above) and two dromedary camels (below) were purchased from Leon Leopard of Lorena, Texas, to become part of the children's zoo in August 1974. C. B. "Bird Dog" Rogers hoped to breed these new animals and then be able to exchange the offspring with other zoos. Watching the antics of these rambunctious individuals delighted young and old alike as they wondered through their ever-changing zoo. (Courtesy of Marsha T. Lautenbach.)

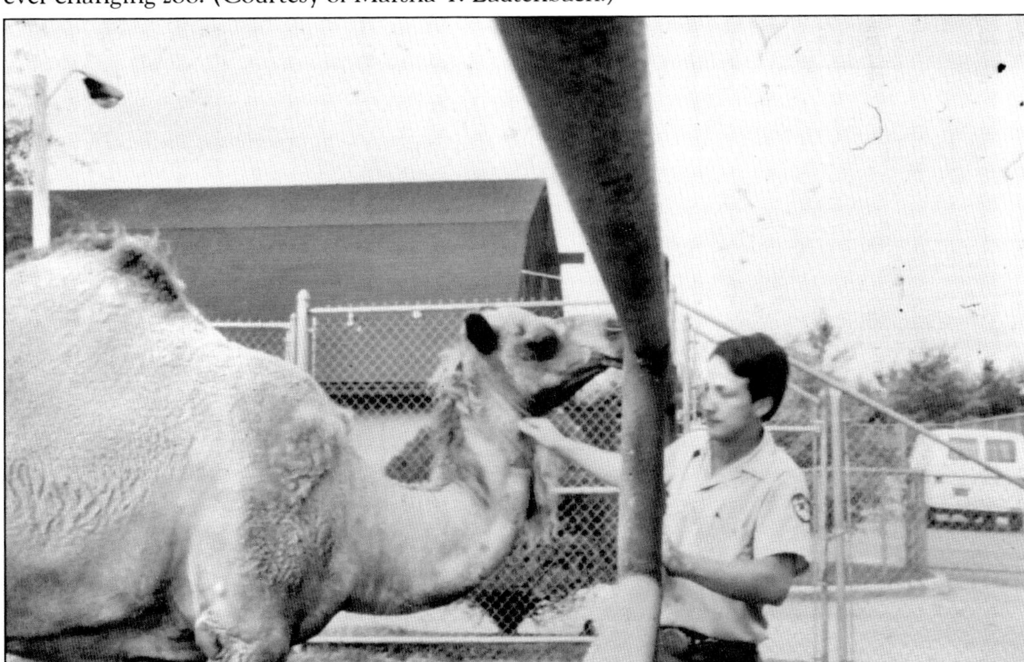

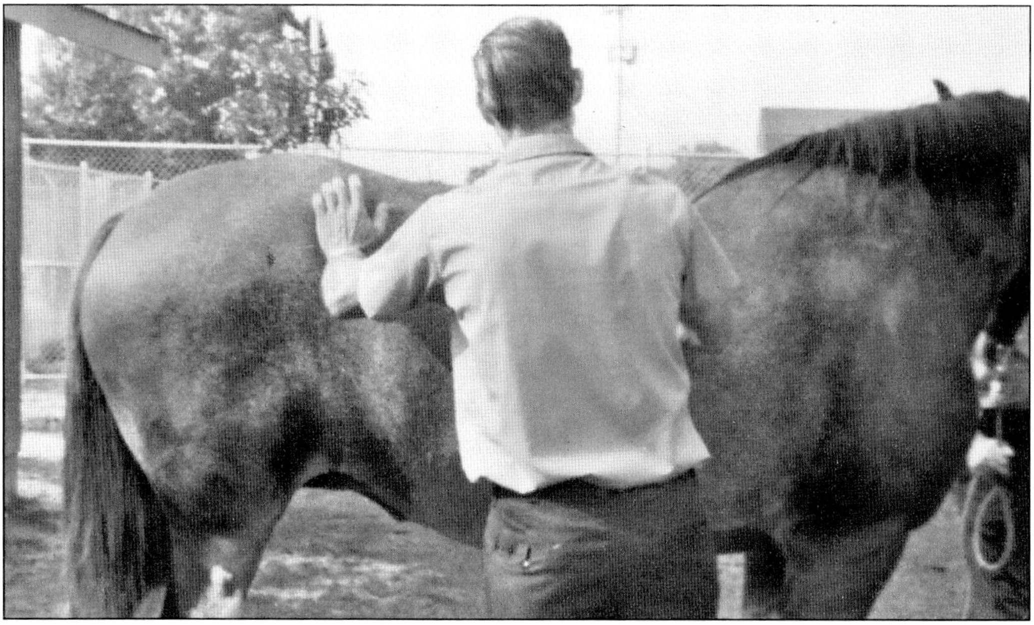

Drumless Shamrock, a 12-year-old female Clydesdale weighing 1,800 pounds, was shipped to Montgomery from St. Louis's Busch Gardens in July 1974. Arrangements for the horse took more than a year, since the city had to provide an adequate home before shipment. She was the first of several Cyldesdales the St. Louis firm intended on donating to U.S. zoos. (Courtesy of Montgomery Zoo.)

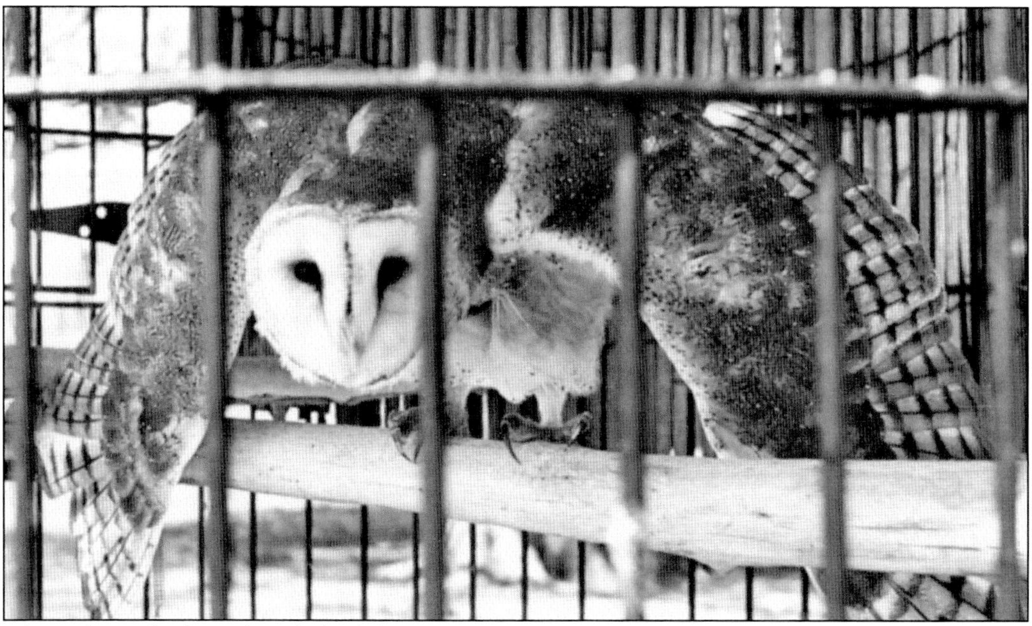

Along the paths from the Clydesdale were circus-type cages that housed a variety of mismatched animals, such as owls, skunks, pangolins, armadillos, hedgehogs, and kinkajous. It was clear that this small children's zoo was growing, and organization of the animal collection and houses was eminent. Even this barn owl could smell the scent of change in the air. (Courtesy of Richard Hammond.)

Three

Montgomery Zoo 1974–1990

In July 1974, the zoo marked the passing of its adolescence when the parks and recreation board approved a name change to the Montgomery Zoo. C. B. "Bird Dog" Rogers, the third zoo director, added four new exhibits, publicized the zoo by television and newspaper media, aroused public interest in the zoo by speaking to many civic and service clubs, and enlarged the zoo staff.

Herbert Rice, president of the Montgomery Area Zoological Society, appointed a long-range planning committee to investigate the feasibility of a master plan and additional acreage for the zoo. Land was made available by the city, with a master plan to develop the additional 33 acres and the existing zoo.

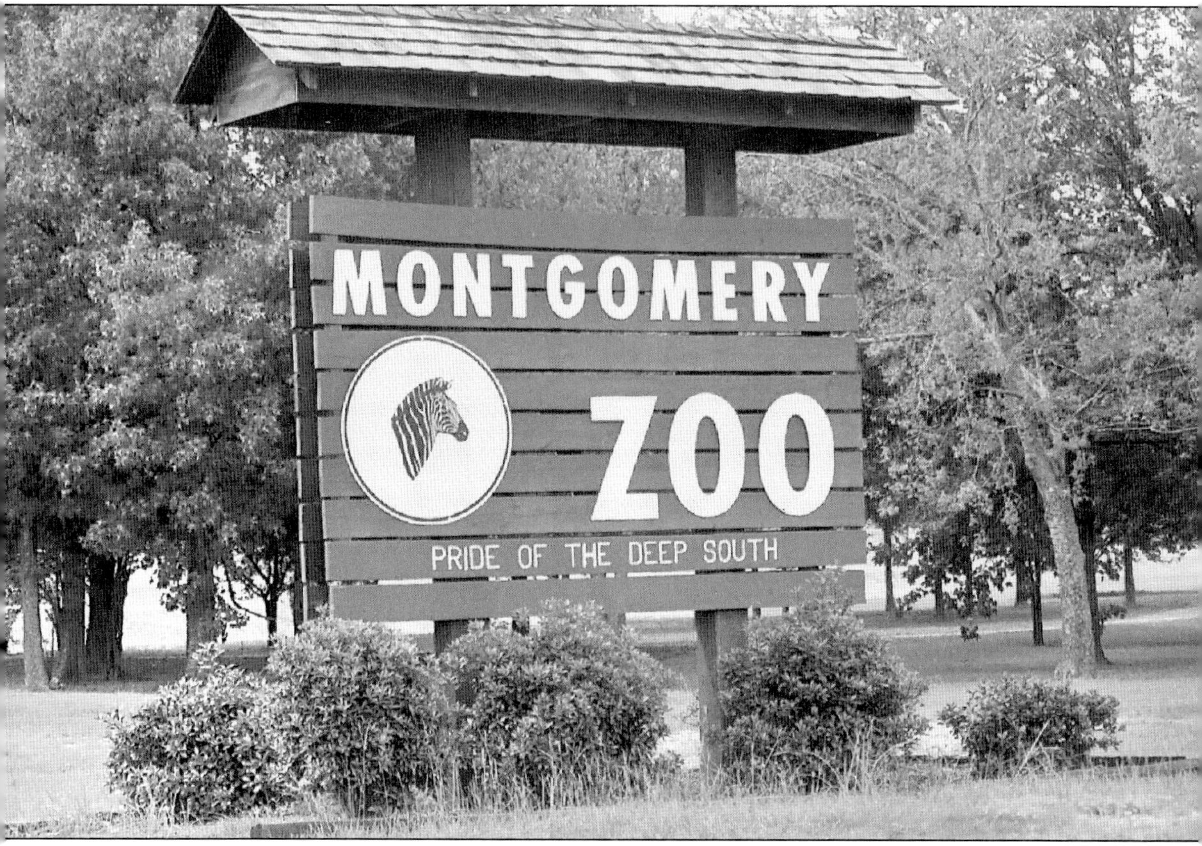

With a name change approved, the Montgomery Children's Zoo became the Montgomery Zoo and therefore needed to modify its animal collection to better represent the main Montgomery Zoo status that it just inherited. C. B. "Bird Dog" Rogers had plans in the works for this new expansion. The new name conveyed to the public of Montgomery that their zoo would now become more than just a petting zoo for children; it would now be an educational and recreational facility that all members of the family could enjoy together. Along with updating several existing exhibits, exhibits for new species were planned. In order to make these plans a reality, funding possibilities were examined and started. (Courtesy of Marsha T. Lautenbach.)

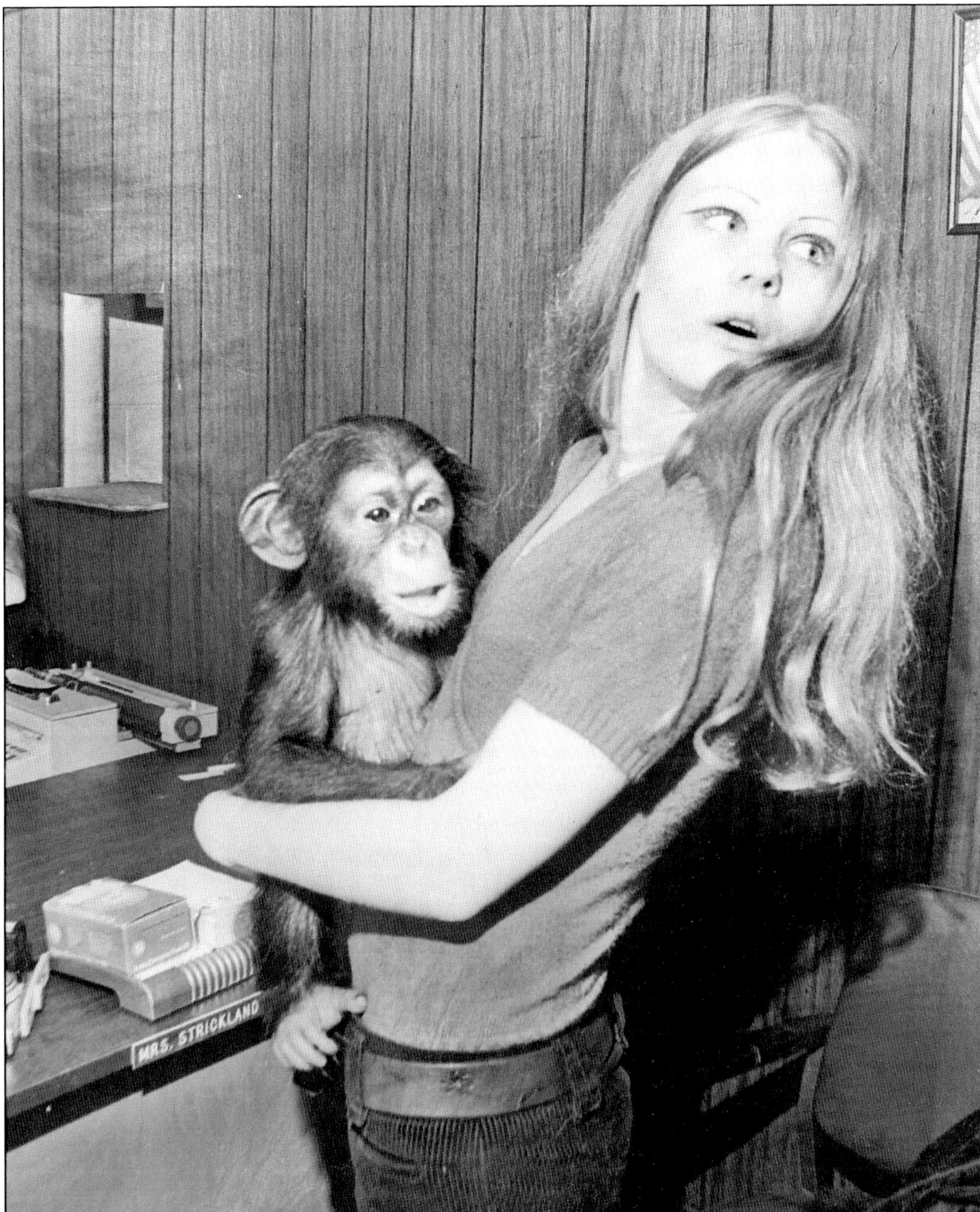

The locally famous chimpanzees Liza Jane, now 45 pounds, and Benji, now 60 pounds, were growing up fast and quickly outgrowing their current 9-by-12-foot enclosure. Zoo personal were starting to fear that the two chimps would be out and about the zoo one day when they arrived in the morning if the exhibit was not updated soon. By March 1975, it was obvious that the zoo needed to make plans for a new, larger exhibit fit for two adult chimps or face the reality that they may need to send their beloved chimps to a facility that could properly and safely display them. Unfortunately the funding for this new exhibit was not forthcoming, and plans to trade Benji and Liza Jane for smaller animals that could be exhibited in the present enclosure had begun. (Courtesy of Montgomery Zoo.)

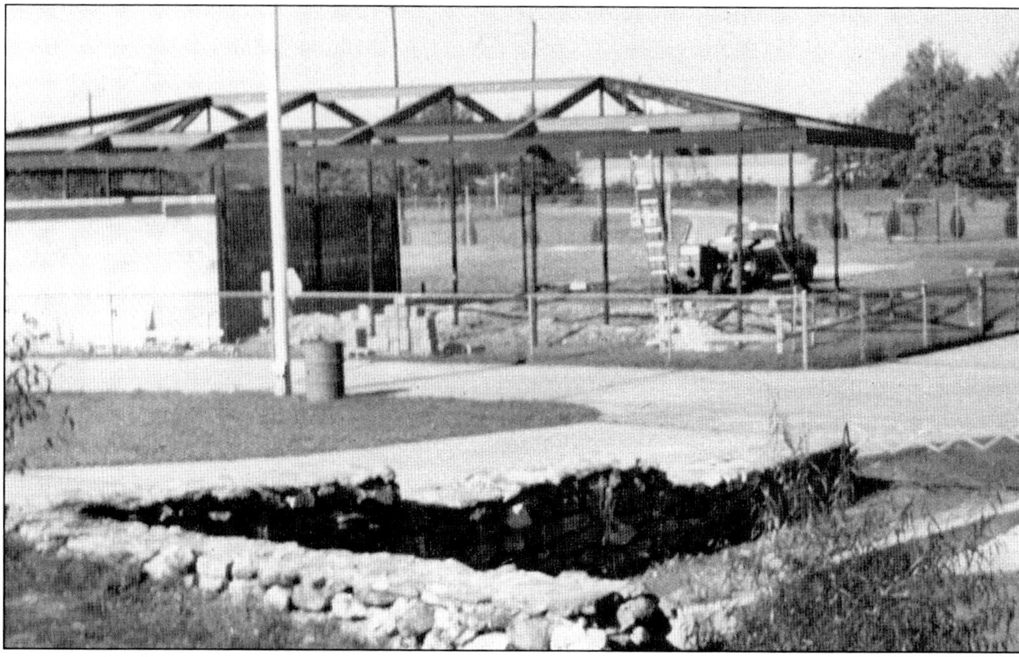

Despite the fact that Rogers was told by the parks and recreation department of Montgomery that there was no money available to build a new chimp exhibit, he continued to hope and look for money from other sources. To build a new chimp facility that met the federal minimum standards would cost a minimum of $7,000. In April 1975, after inquiries with the city controller's office, Rogers discovered a $15,000 "veterinary services" fund that could be used to build better facilities. Deciding to use part of this fund to build the new enclosure, Rogers announced to the public that the chimps would be staying in their hometown of Montgomery. (Courtesy of Montgomery Zoo.)

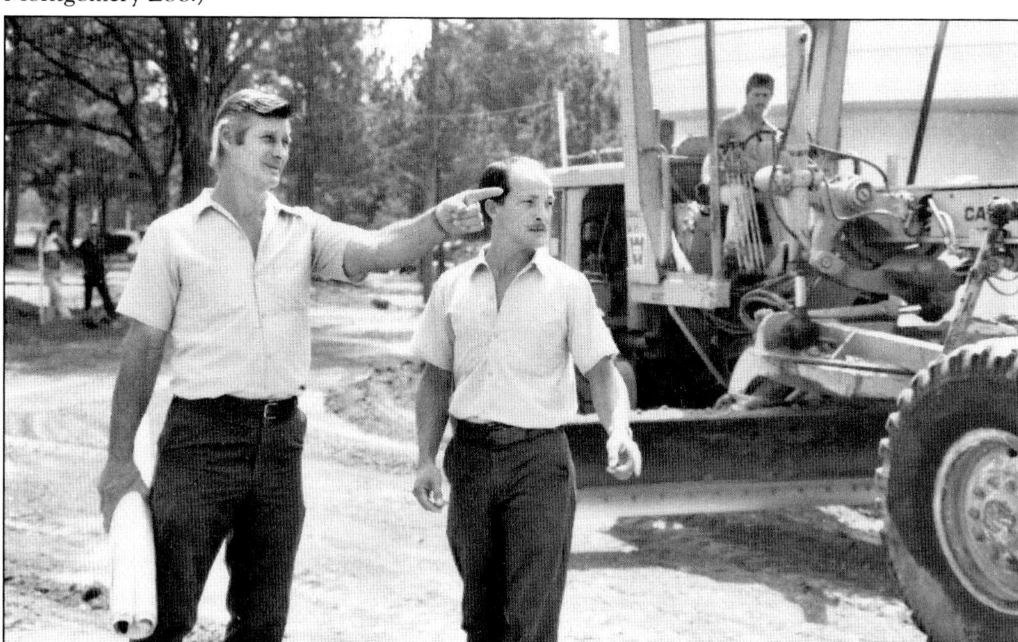

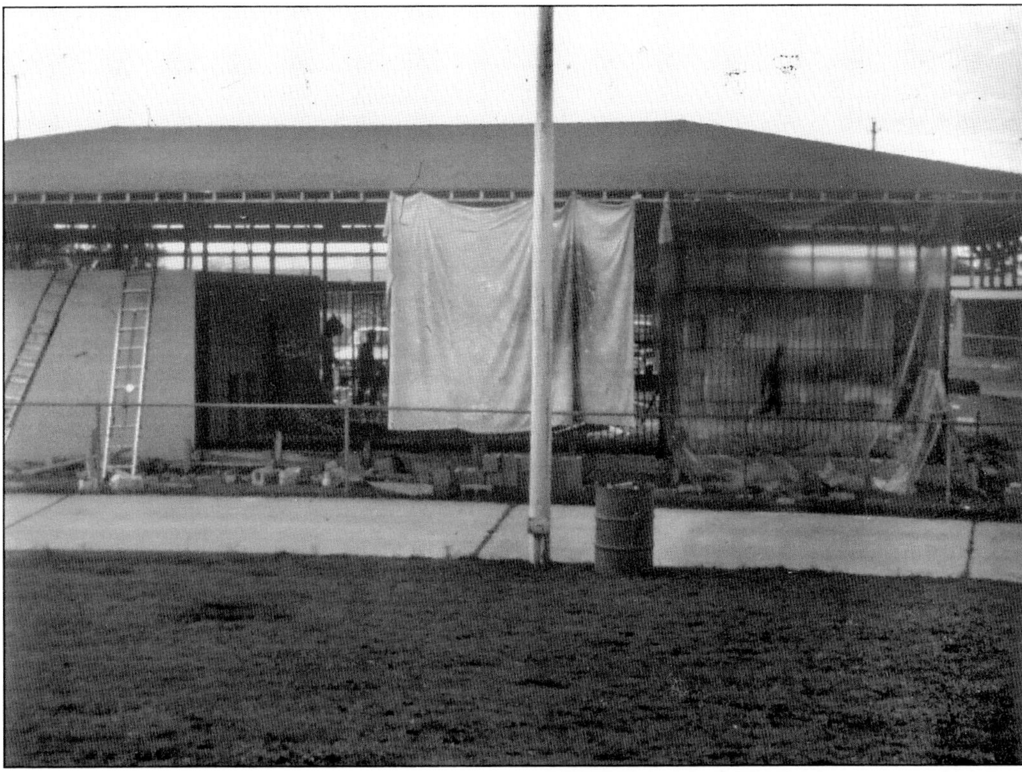

With funding obtained, plans and construction for the new chimp enclosure had begun. Rogers designed a 25-by-35-by-8-foot outside area with steal bars and climbing structures. The new facility also included three indoor bedrooms that were electrically heated during the winter. For Benji and Liza Jane, the exhibit could not be finished too soon. The month before the exhibit was to be completed, the chimps both developed infections when the temperatures unexpectedly dropped. They were put on antibiotics to help with their sniffles and recovered quickly, but the new heated bedroom areas were eagerly awaited by the two. (Courtesy of Montgomery Zoo.)

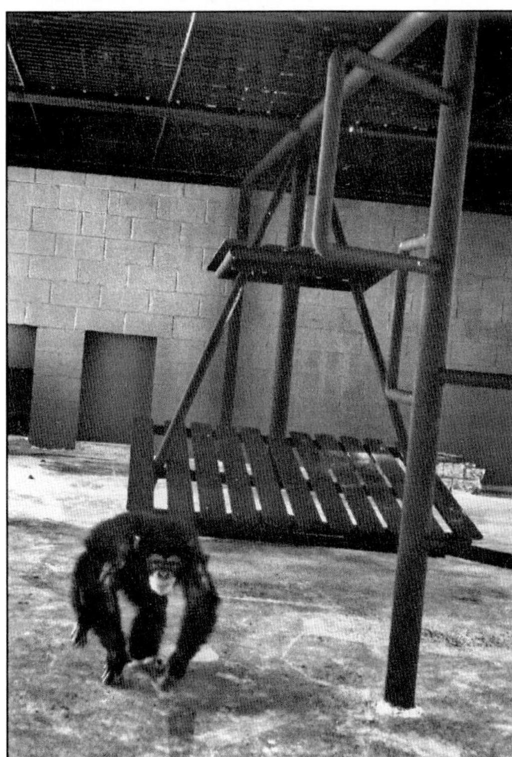

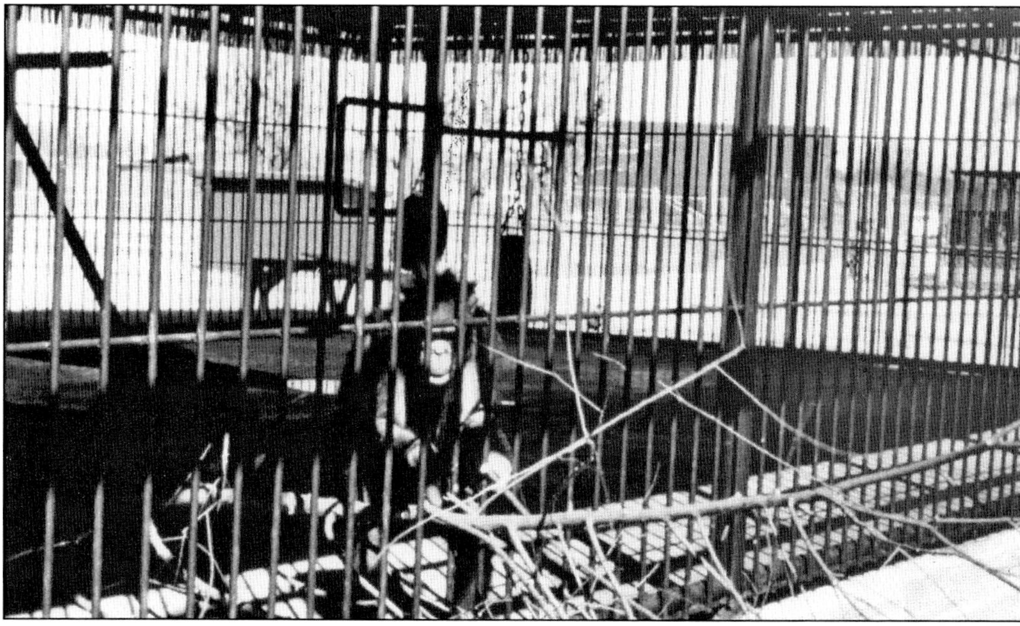

Benji and Liza Jane welcomed their new home on November 1, 1975. The new chimpanzee enclosure was the largest display structure at the Montgomery Zoo. But the new home ended up costing $26,551, well over the expected $7,000, using the entire veterinary services fund and extra money being provided by the City of Montgomery. This would leave no city money for other projects at the zoo. (Courtesy of Montgomery Zoo.)

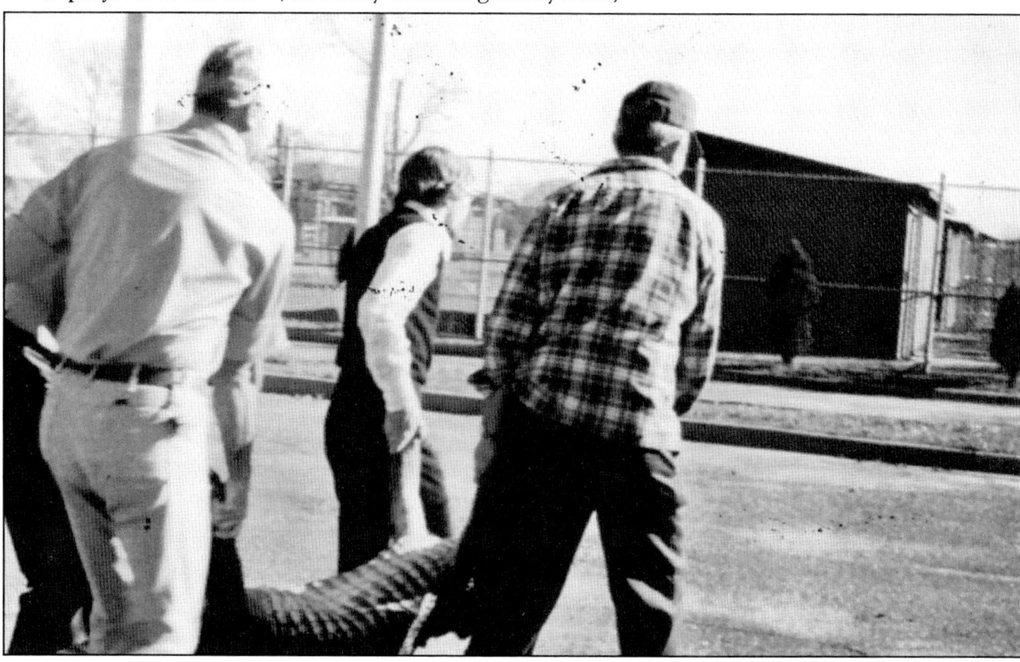

In conjunction with the chimp exhibit, construction of a new alligator exhibit had also begun. The seven alligators that called the Montgomery Zoo home got a new exhibit in December 1975. The alligators were moved from the concrete moat surrounding Monkey Island to their new grassy area across from the chimps. (Courtesy of Montgomery Zoo.)

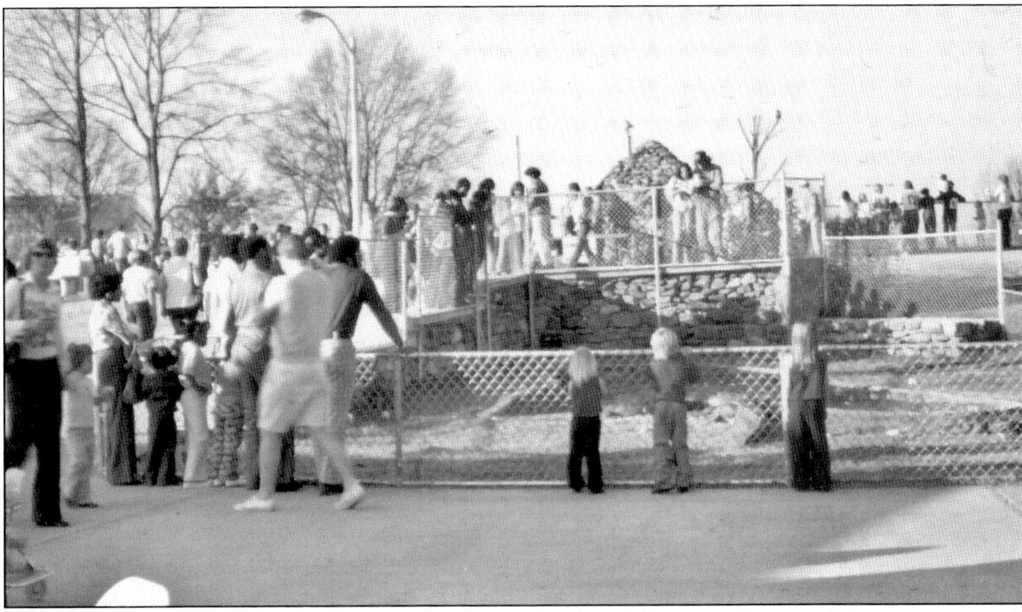

In the new enclosure, located between Monkey Island and the new chimp exhibit, visitors could view the alligators and alligator snapping turtles up close from the level of the alligators or from a platform positioned above the exhibit. Though the alligators no longer had views of the monkeys on the island, the largest alligator, Number One, and his six smaller fellow alligators could now spread out on the grass of their larger enclosure. Basking together in the sun, the alligators were now teased by the peacocks that walked around the zoo. Guests would say that it looked like Number One was hoping one would pop in the enclosure for a one-time visit, but the peacocks always stayed just out of reach on the other side of the fence. (Courtesy of Montgomery Zoo.)

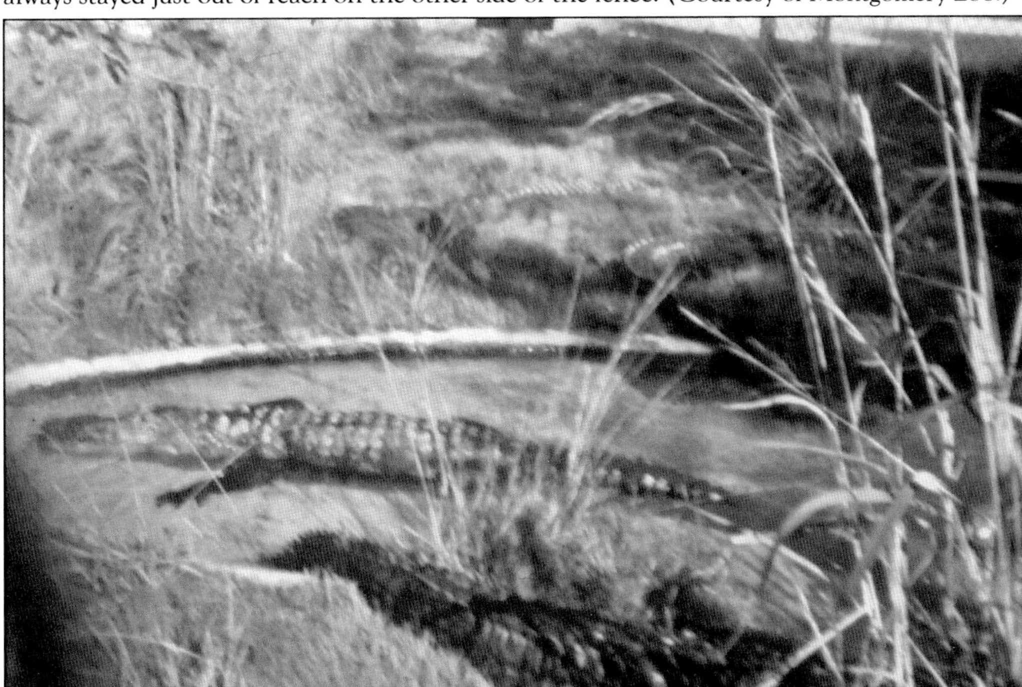

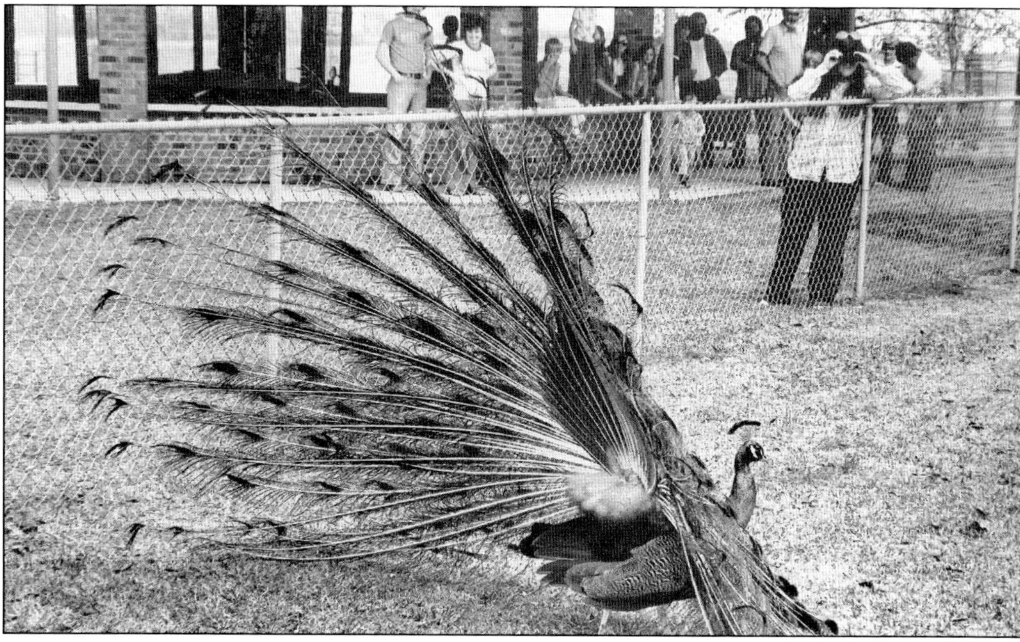

Finally completed in 1975, three years after ground-breaking, the Safari Shelter was a covered educational center. At the center of the structure was an enclosed area that housed reptiles in the summer and birds in the winter. Surrounding the center exhibit were enclosed pens that contained tractable animals that educators could bring up to the window areas. This allowed the public to interact with the animals and ask questions of the educators. This new approach to animal housing allowed visitors the unique experience of interacting with animals other than the usual petting zoo domestics while also learning at the same time. The construction of the Safari Shelter was sponsored completely by the Capitol Junior Women's Club and Naval Reserve members. (Courtesy of Montgomery Zoo.)

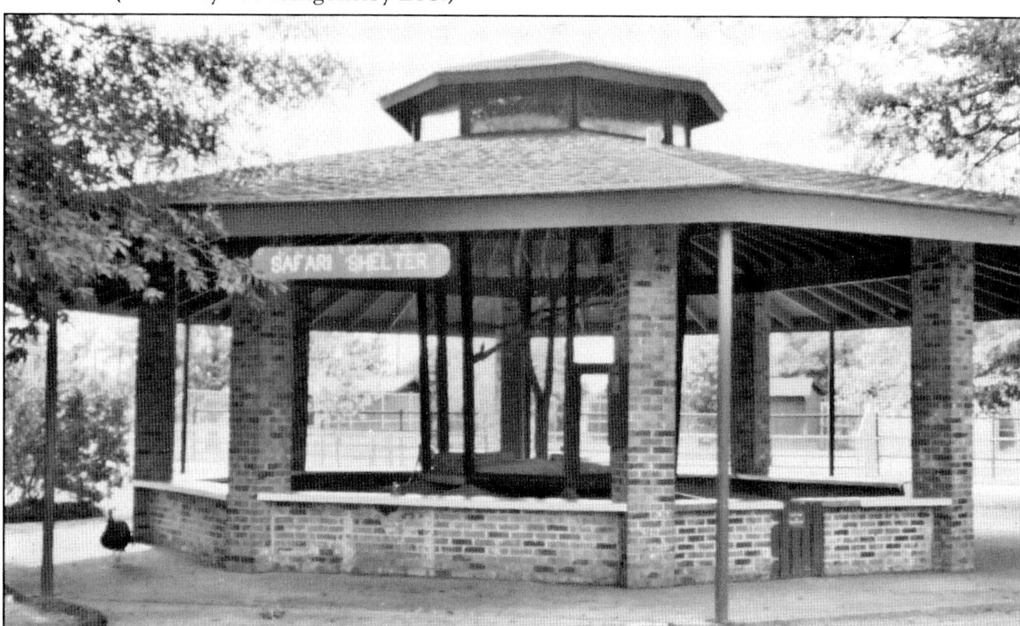

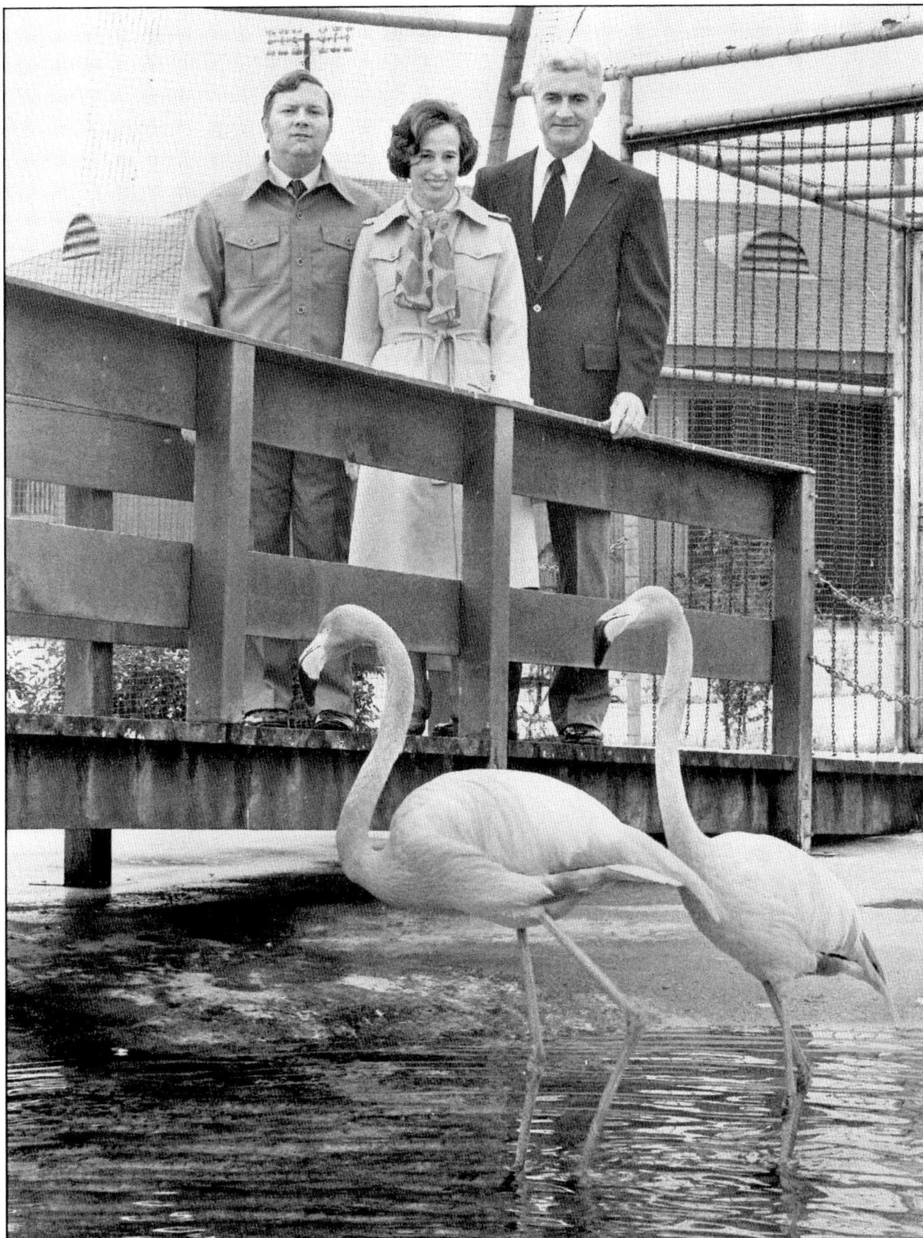

The Dixie Zoological Society, Inc., was a non-profit organization separate from the Montgomery Zoo but worked with the zoo to support the facility for the benefit of both the animals and the visitors. As a non-profit organization, the Dixie Zoological Society (DZS) was able to raise money that it could then allocate to the conservation and research needs of the zoo as it saw fit. The more visible activities of the society included the publishing of a monthly newsletter, organizing fund-raising events, and volunteering at zoo functions. The society was instrumental in the zoo's growth, providing both the funds and volunteer support for many of the new additions to the zoo. Above, Larry Withers, DZS president (left); Martha McInnis, DZS vice president (center); and C. B. "Bird Dog" Rogers, zoo director, posed for a photograph in the flight cage. (Courtesy of Montgomery Zoo.)

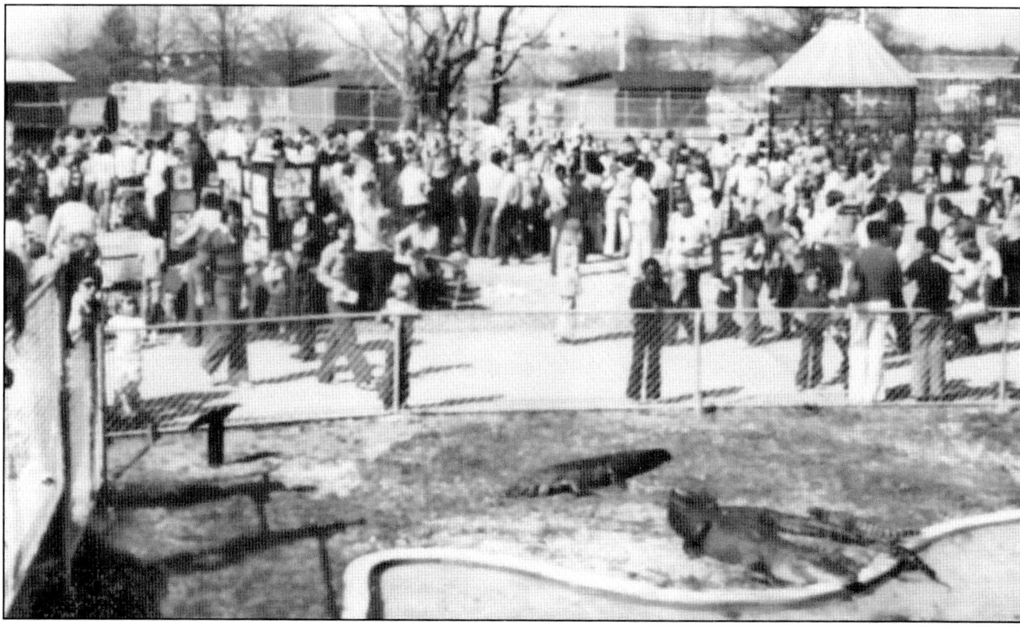

In December 1975, Ron Mason approached zoo director Rogers about an idea for a promotional event between the then-new 102-FM WHHY station and McDonald's to help the Montgomery Zoo. This event became named "Zoo Day," and on February 14, 1976, over 16,000 visitors attended the free events at the zoo. The Zoo Day event became a yearly fund-raiser for the zoo. (Courtesy of Montgomery Zoo.)

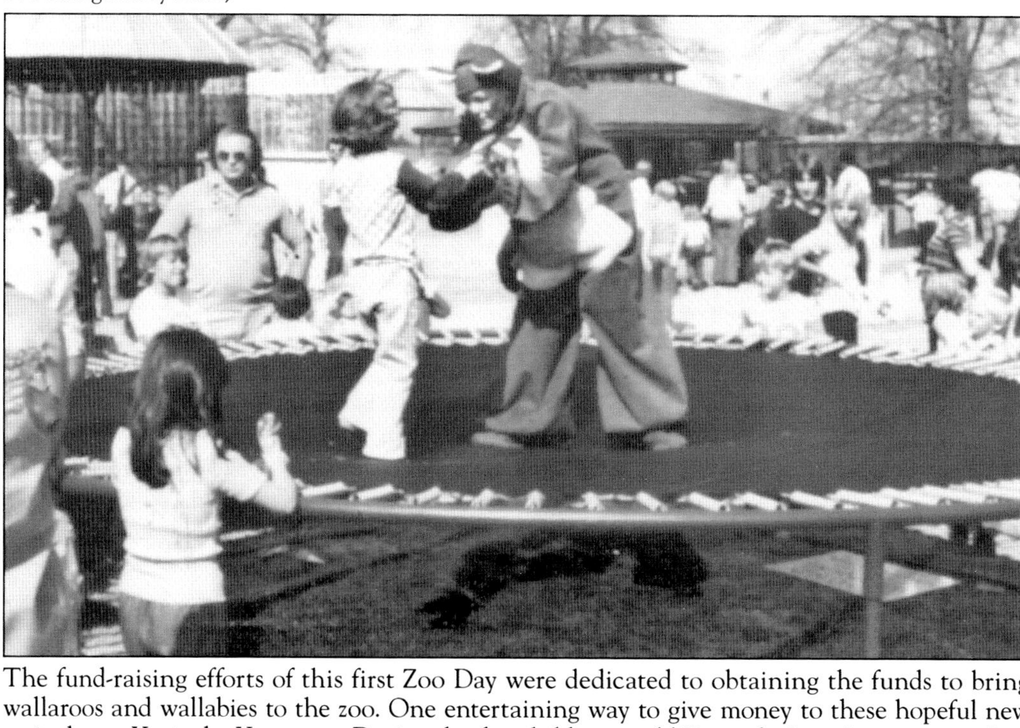

The fund-raising efforts of this first Zoo Day were dedicated to obtaining the funds to bring wallaroos and wallabies to the zoo. One entertaining way to give money to these hopeful new arrivals was Katie the Kangaroo. During the day children could "Jump for Joeys" on a trampoline with Katie. (Courtesy of Montgomery Zoo.)

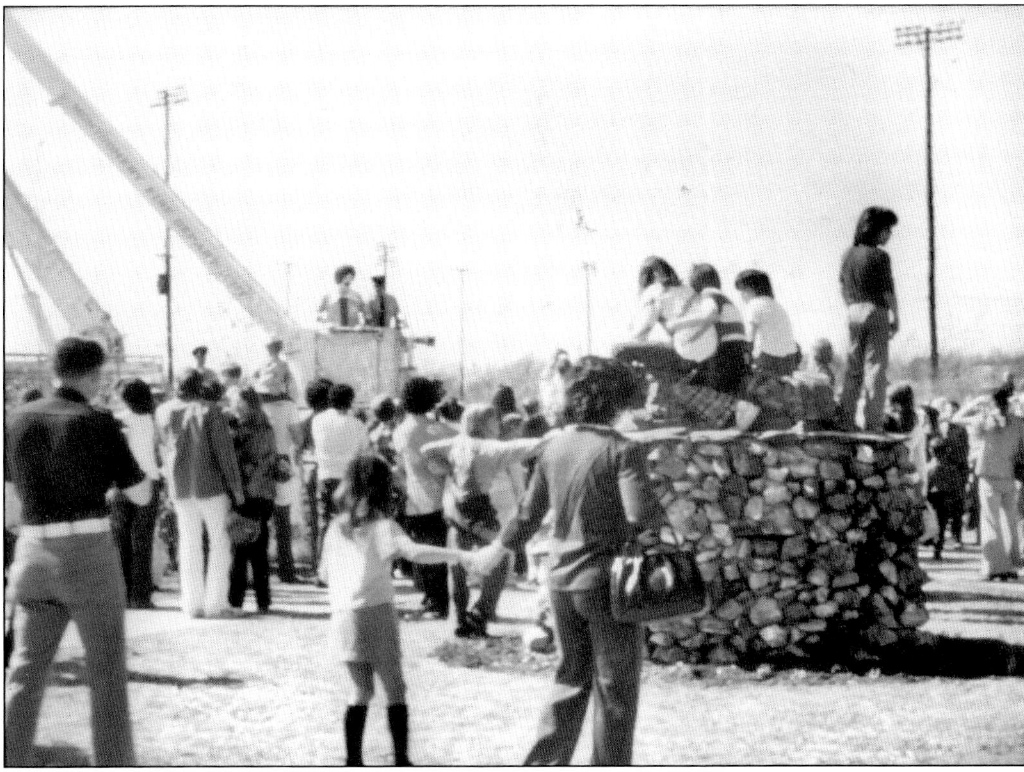

As part of the Zoo Day special entertainment, Ronald McDonald arrived at the zoo via the Montgomery city fire department's snorkle. After this mesmerizing entrance, he and his McDonaldland friends preformed two shows during the day, much to the excitement of the younger visitors. (Courtesy of Montgomery Zoo.)

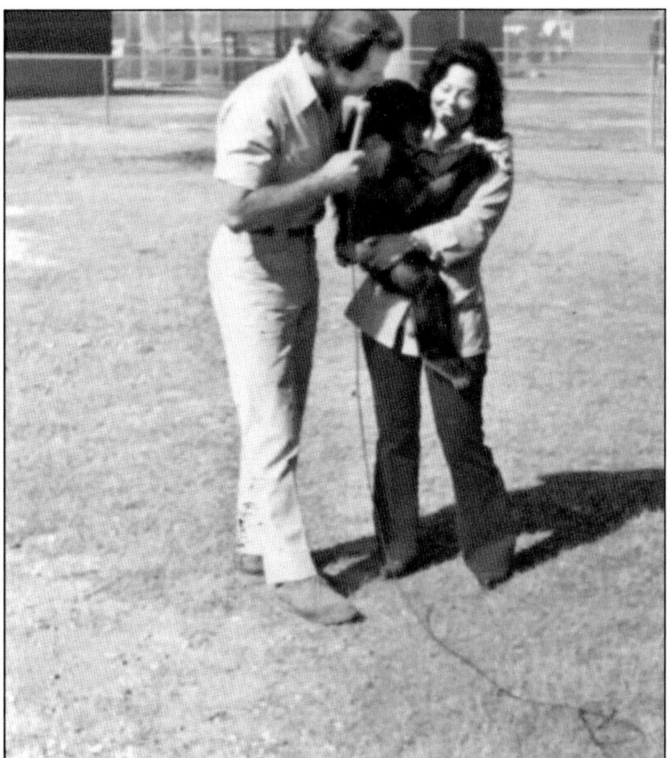

Stan Brock from *Mutual of Omaha's Wild Kingdom* made a guest appearance at the Montgomery Zoo's Zoo Day. While at the event, he made a public service announcement spot promoting the Dixie Zoological Society. Stan Brock appeared with Benji, the chimp, in the park and for television interviews. (Courtesy of Montgomery Zoo.)

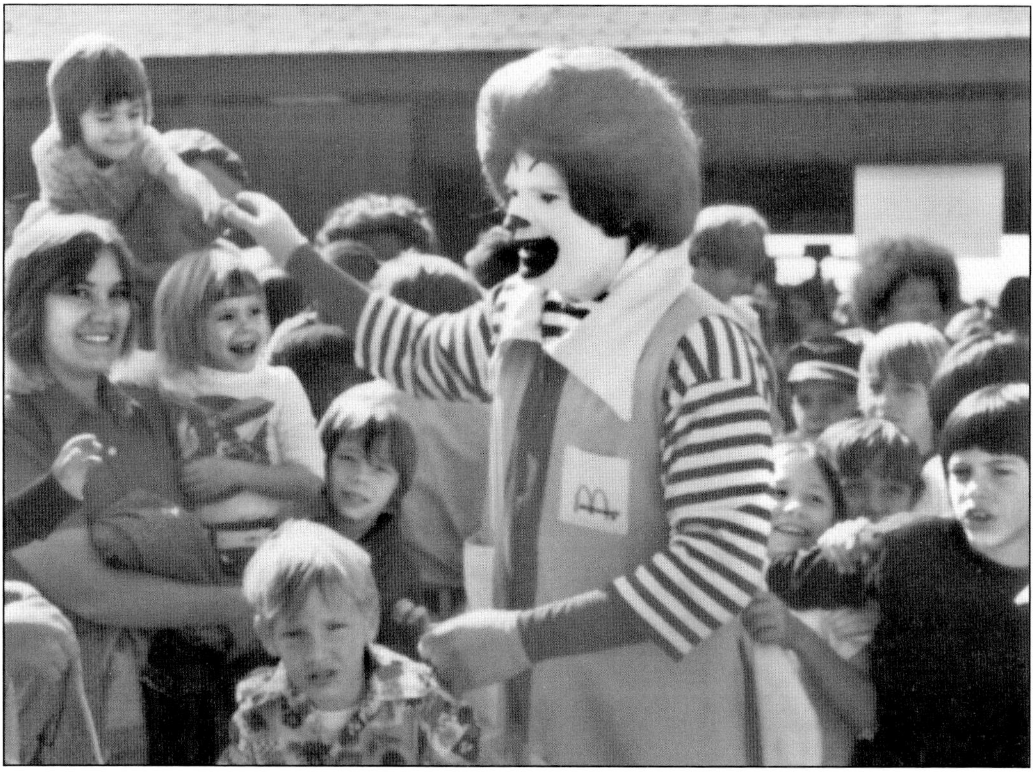

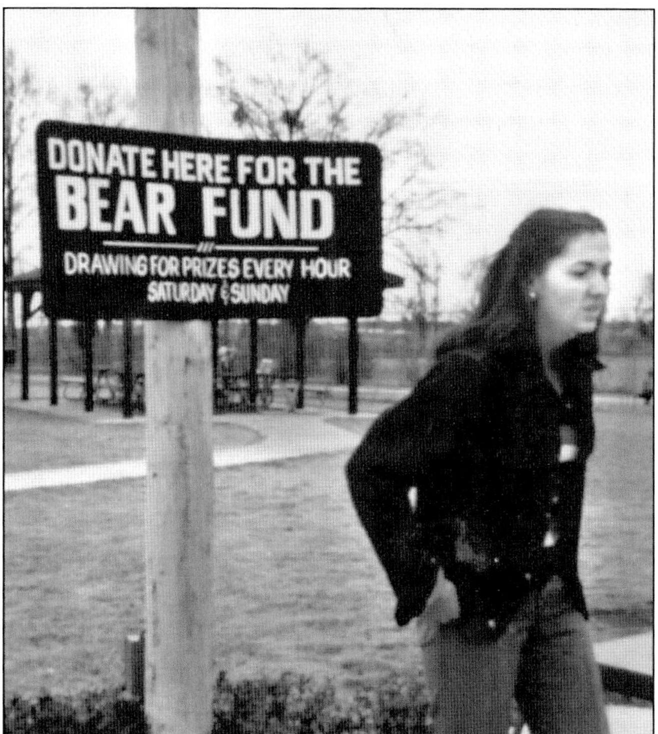

WHHY, a local radio station and sponsor of Zoo Day, broadcasted live from Monkey Island. Radio personalities, while soliciting funds to bring the wallaroos and wallabies to the zoo, invited citizens of Montgomery to the zoo and see for themselves the many entertaining activities going on for this special day. (Courtesy of Montgomery Zoo.)

Though the focus of this first Zoo Day was to raise money for kangaroos, other fund-raising events for bears were held during Zoo Day in February 1976. Signs were posted around the zoo, with special areas for visitors to make donations to the bear fund. Prizes were given out every hour on the weekends. (Courtesy of Montgomery Zoo.)

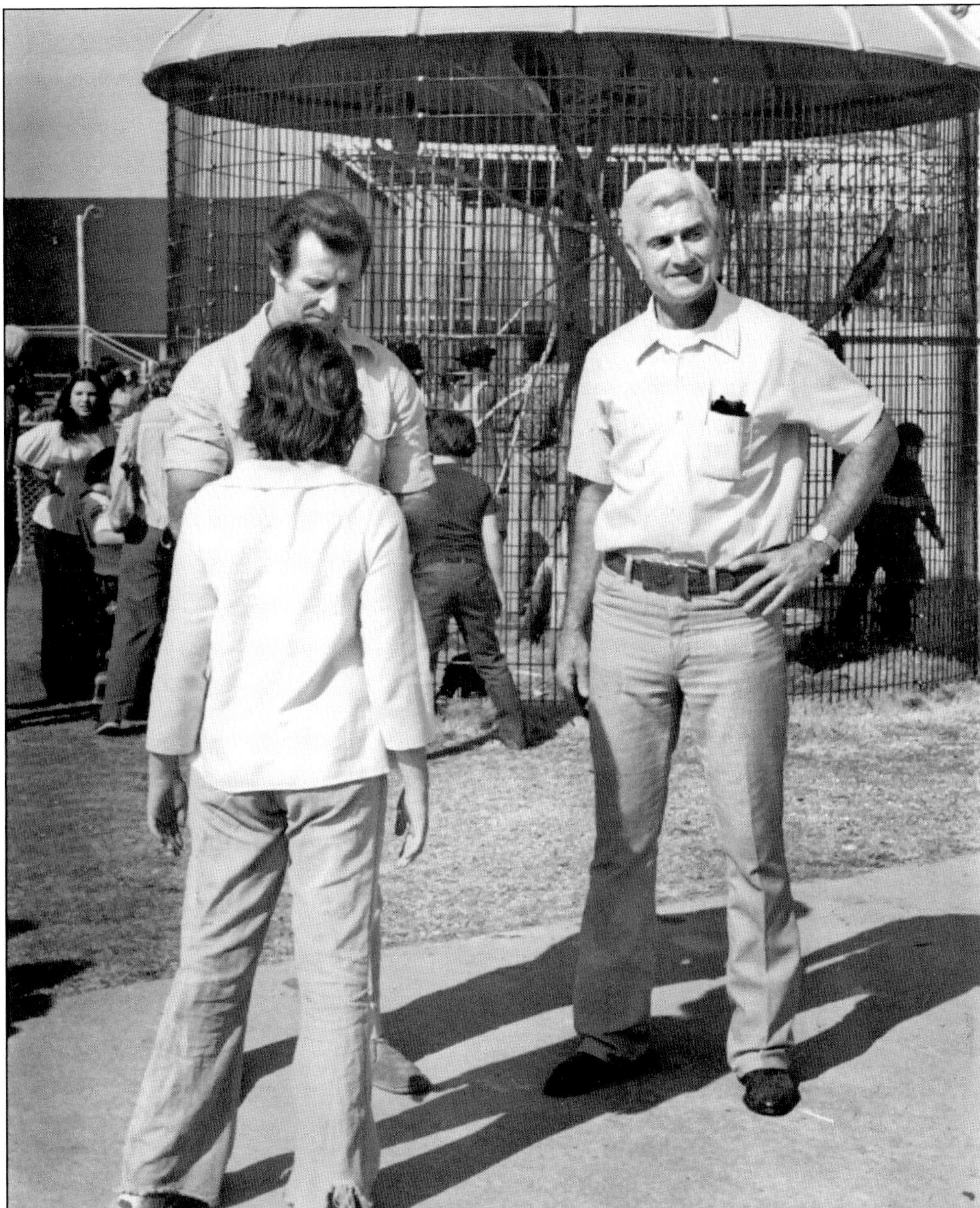

While being the main guest for Zoo Day, Stan Brock toured the zoo with director C. B. "Bird Dog" Rogers. This tour gave the visitors an opportunity to interact with a television legend and talk about their local zoo with its director. By being able to bring a big-name celebrity to its main fund-raising event, the Montgomery Zoo was able to increase attendance and proceeds to benefit the zoo's hopeful addition of wallabies and wallaroos. His visit to the zoo had an obvious impact on his impression of Montgomery, because *Mutual of Omaha's Wild Kingdom* would have more involvement with the zoo in later years. (Courtesy of Karen W. Carden.)

Zebra Tales

Official Newsletter of the DIXIE ZOOLOGICAL SOCIETY

Volume 1, Number 4 July–August 1978

ZOO DIRECTIONS

Bob Cooper, Zoo Director

On the evening of June 27, Mr. Sabin Robbins, Executive Director of the Friends of the National Zoo, spoke before some of the zoological society members. This was the first membership meeting that the society has had since I came to work at the Montgomery Zoo a year ago.

The people that attended the meeting were treated to an excellent film about the national zoo and its operations. Mr. Robbins turned out to be a humorous and informative speaker. The sheer massiveness of a zoo operation of that size is overwhelming.

As an added attraction, Barbara Keel and Al Carroway displayed some of their wildlife art. Both are excellent artists. Mr. Carroway is teaching the art class at the zoo during Zoo Summer 78.

Although the turn out was good, many members were not present. Hopefully, they will come to our next meeting. Martin Darity, program chairman, is sure to have another excellent program planned.

These programs are just a part of the fun of being a member of the Dixie Zoological Society. Membership entitles you to free admission to the Montgomery Zoo and many zoos throughout the country, and a subscription to ZEBRA TALES. I suppose you have noticed the new format for the newsletter. Let us know how you like it.

Everyone will be glad to know that the Dixie Zoological Society now has enough money to build a bear exhibit. The zoo staff is finishing the design and the cost estimate for materials. The Montgomery Parks Department will provide labor. If we are lucky, the new exhibit will be completed by Zoo Day 1979. The money for this project was raised on past Zoo Days that are organized by WHHY and McDonald's Hamburgers. This is a true community effort. Next Zoo Day funds will go for animals that can live in existing exhibits.

The two year old Dixie Zoological Society is certainly becoming a worthwhile organization that benefits the members and the zoo. A zoo is only as good as the zoo society, and the Montgomery Zoo serves all of South Alabama.

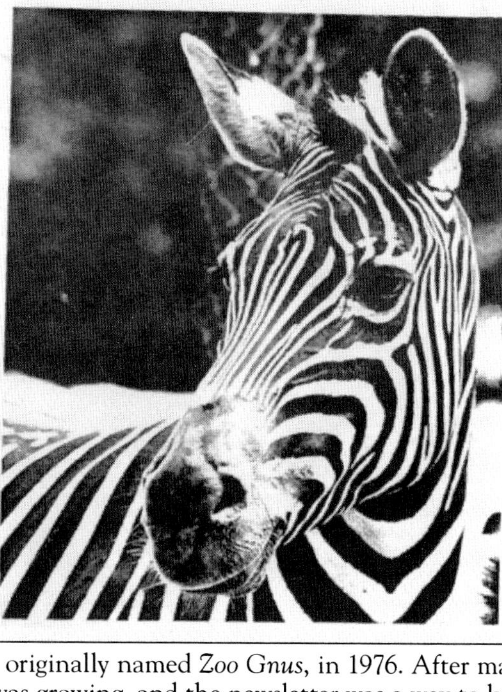

The DZS began a monthly newsletter, originally named *Zoo Gnus*, in 1976. After many drives, membership in the zoological society was growing, and the newsletter was a way to keep all the members up to date with the activities of the zoo. Each issue would contain summaries from both the director and the animal curator of the activities at the zoo since the last issue, an update on the society's projects, and an educational article about one of the many species found at the zoo. With membership in the society, one would receive not only a subscription to the newsletter but also free admission to the Montgomery Zoo and many zoos throughout the country. This newsletter was renamed *Zebra Tales* in 1978 to coincide with the zoo's mascot, a zebra named Zoobie-Doo. (Courtesy of Montgomery Zoo.)

Who's Who in the Zoo

BIRD DOG ROGERS
Montgomery Zoo Director

Stunning to behold, vile-tempered and stubborn, the zebra is a member of the horse family. Because of their stripes, the Romans called them "horse-tigers".

Thair stripes are like human fingerprints, no two are alike and there may also be a very wide variation in any single herd. The zebras displayed here in the Montgomery Zoo are native to a very wide range in Africa south of the Sahara. As a whole, these animals are vertically striped in front, horizontally on the back legs, and diagonally on the rump and flanks. This causes a V-shaped junction pattern about the middle of the sides.

Zebras are typically horse-like grazers and go about in large herds, affording the lion their main article of diet. Zebras are shy and nervous, but are also very pugnacious and can defend themselves with considerable vigor, kicking both backwards with their hind feet and forwards with the front and biting savagely.

Many creatures are color-blind and see the world as though it were an untinted photograph, in greys, blacks and whites. But even a color-blind enemy can be deceived by the kind of coloration called disruptive. This is the marking or mottling that breaks up an otherwise identifiable outline. Zebras color patterns render them extremely conspicious against green backgrounds, but virtually invisible in tall grasses of their native savannah. The complex striping tends to break-up their outline and provides a protective coloration on the dry plains.

These generally gregarious animals assemble in large herds and are often found mingling in herds of other animals such as wildebeests or even ostriches. Old males are occasionally solitary.

In many ways they appear to be closer to asses than horses, having long ears, strong upright manes, terminally haired tails, narrower hooves, and chestnuts usually confined to the front legs.

Zebras are resistant to common African diseases that horses cannot survive. Therefore, efforts have been made by government and private agencies to demesticate them or to cross-breed them with horses, but they are likewise resistant to domestication, so most of these efforts have been disappointing.

After a gestation period of about a year, zebras rely on their keen eye-sight and sense of smell to avoid danger as they strive to battle their often times harsh environment to live out their life span of 25 or more years.

You may see these striking, pugnacious equines, Zelda and Zorro, who appear to be wearing striped pajamas, every day between 9:30 a.m. and 4:30 p.m. at your Montgomery Zoo.

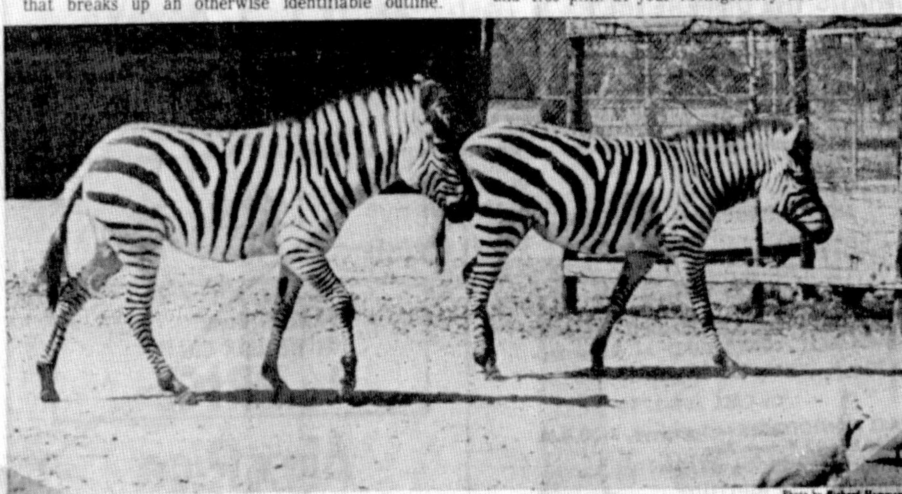

As the new Montgomery Zoo director, one of C. B. "Bird Dog" Rogers's responsibilities was to educate the public of Montgomery and publicize the zoo's growth from a children's zoo. On March 28, 1976, Rogers began writing a weekly column for a local newspaper, the *Montgomery Advertiser/Alabama Journal*, titled "Who's Who at the Zoo." In each of these articles, Rogers focused on a different species that was presently residing at the zoo, providing a photograph of the animal at the zoo, the natural history of the species, and information about the individuals at the zoo. Basic zoo information such as when would be a good time to see the animals and any special events that may be going on that week were also included at the end of each article. Through this column, Rogers was able to both educate the public and encourage visitation to the Montgomery Zoo. (Courtesy of Montgomery Zoo.)

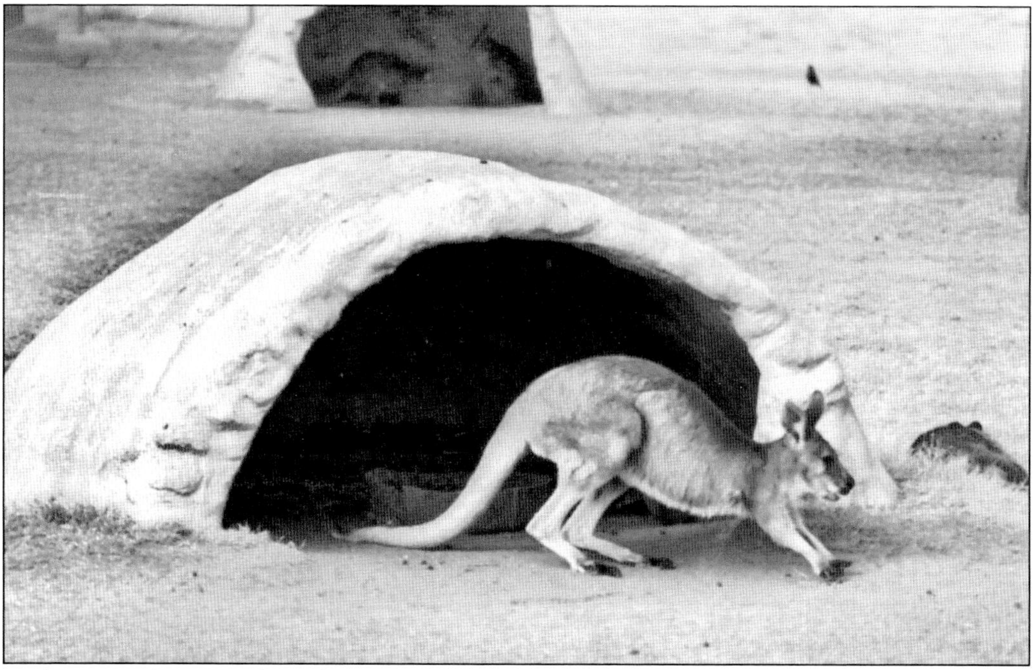

With monies that came through Zoo Day 1976 and Zoo Pals, a Preferred Animal Lovers program in which children and other individuals can make donations to their zoo, the first new species Rogers planned to obtain became a reality. In May 1976, the zoo obtained two wallaroos and two wallabies, which would grow into the large pods they have today. (Courtesy of Marsha T. Lautenbach.)

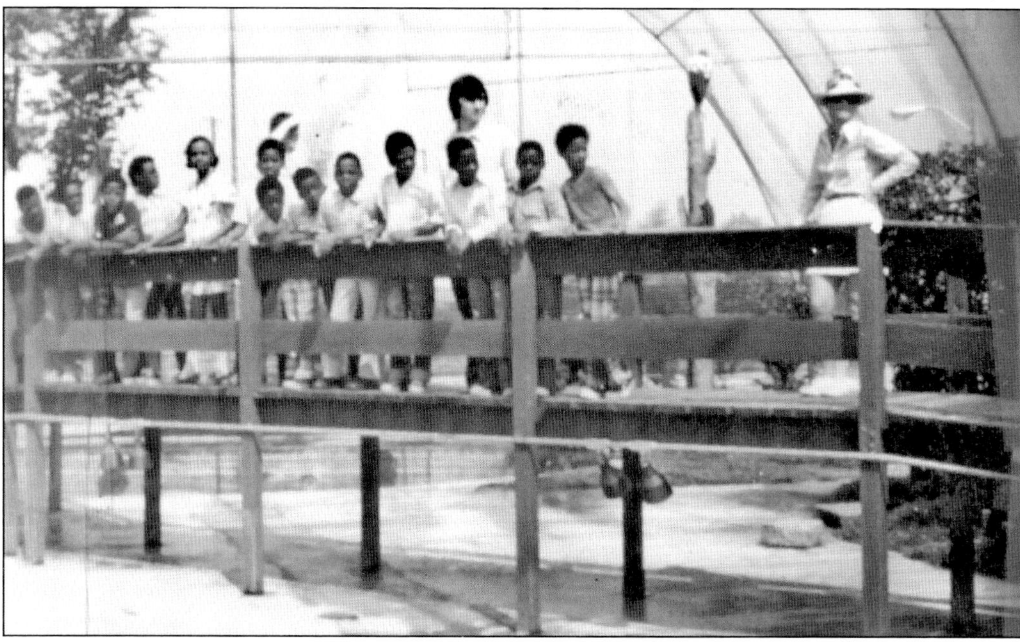

As new species were added, more opportunities arose to educate the zoo-going public. To greet this opportunity, the Montgomery Zoo expanded its education program by adding more zoo tours for children. Classes about different animal-related subjects were led through the zoo by volunteers of the Capitol Junior Women's Club. (Courtesy of Montgomery Zoo.)

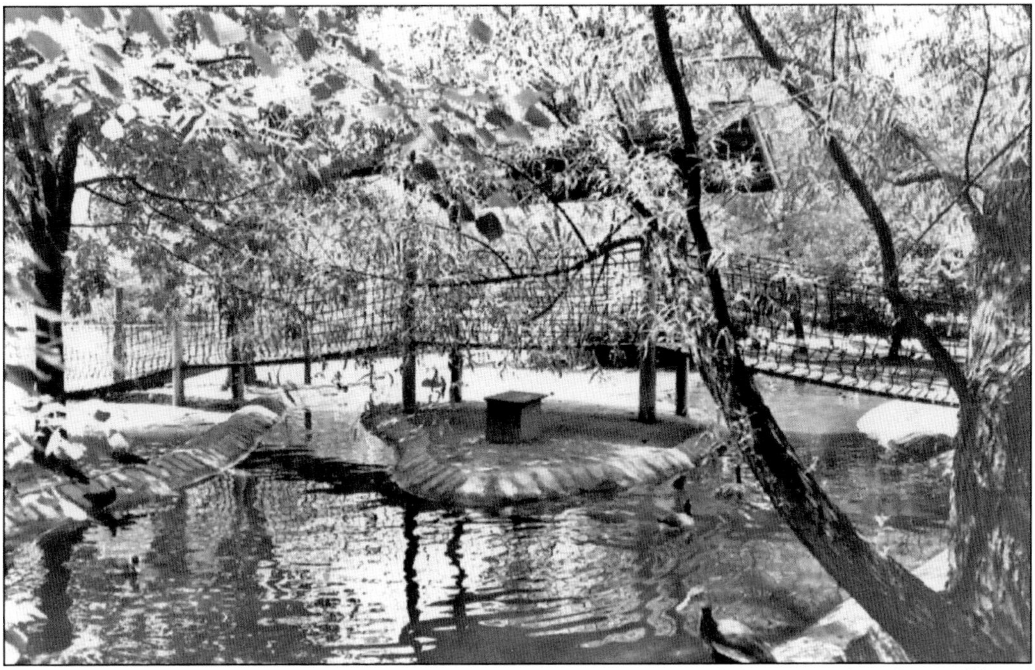

Adding to the zoo's new attractions in July 1976 was a waterfowl exhibit created to show off local and exotic ducks, geese, and swans. Wading birds were also exhibited, along with the various waterfowl. One of the more interesting birds was the mysterious black swan (below), which is native to Australia. The exhibit is an open enclosure that visitors can walk through, following the winding path along the man-made stream to have an up-close encounter with the inhabitants. Quieter moments can be spend just watching the birds go about their day from the wooden bridge that connects the opposite sides of the exhibit. (Courtesy of Marsha T. Lautenbach.)

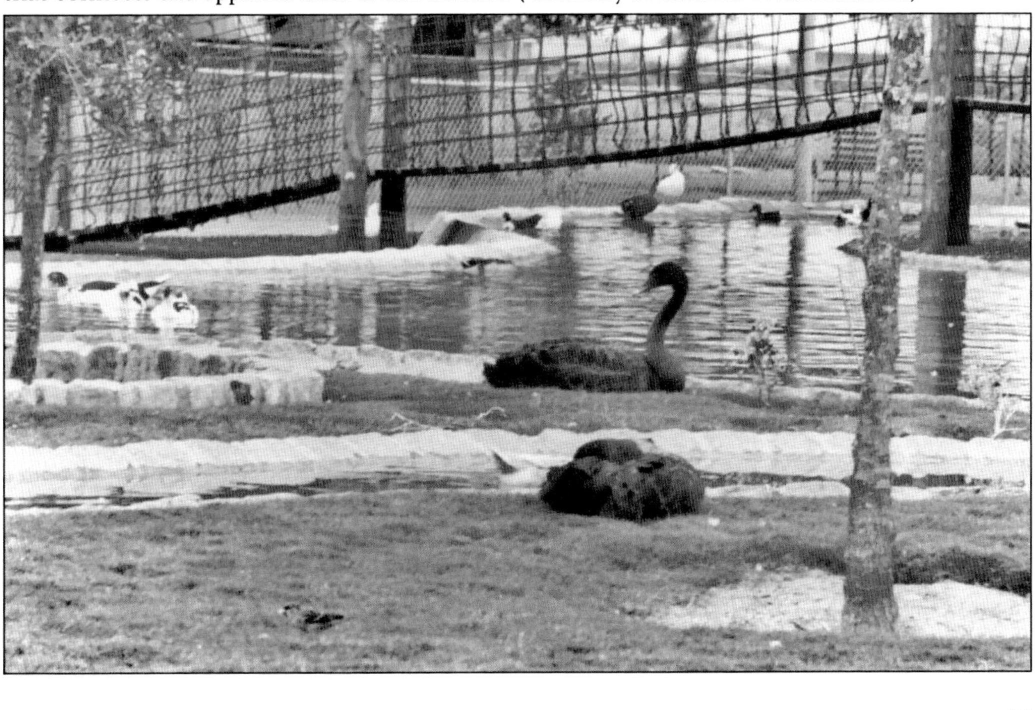

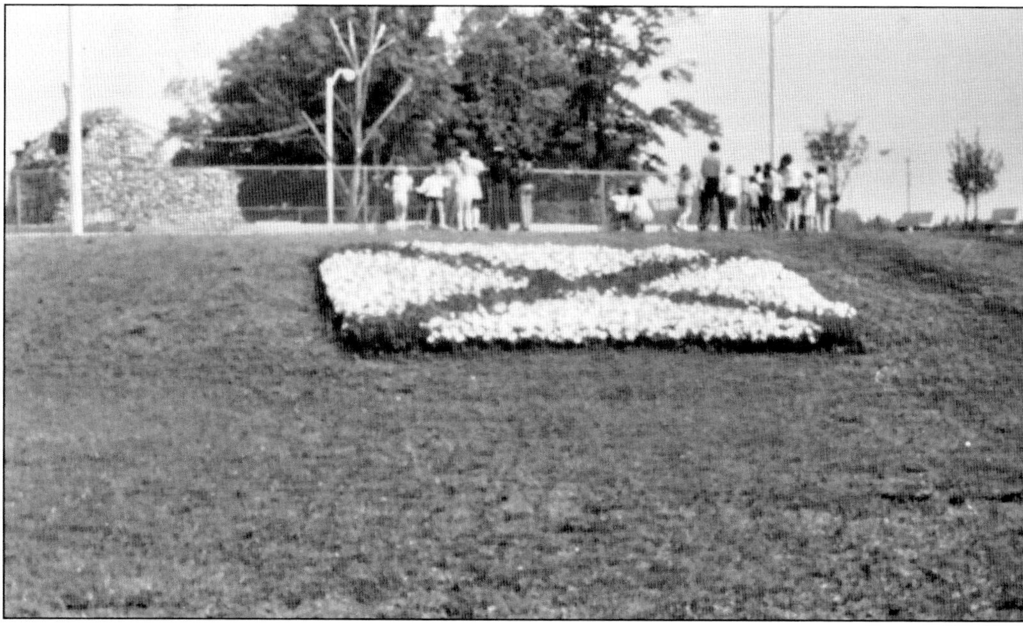

Matt Wendling, a city horticulturist, worked on the zoo grounds to create a botanical as well as a zoological experience for the visitors. One of his more whimsical projects was a 12-by-18-foot flag made of red and white petunias. This floral masterpiece was created for the country's bicentennial and was the only state floral flag in the country. (Courtesy of Montgomery Zoo.)

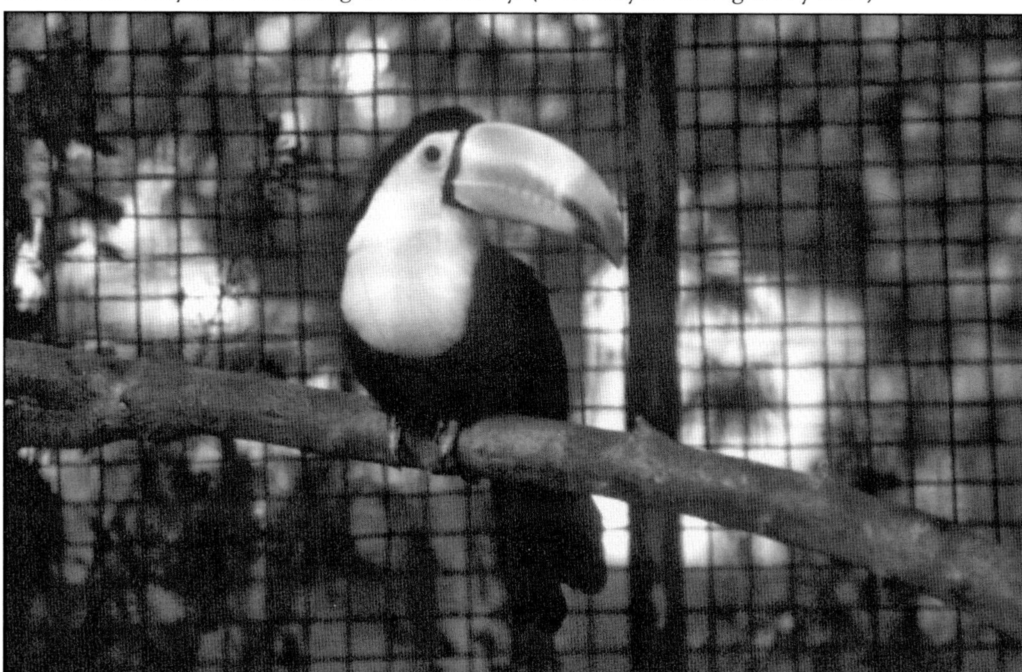

Teeter the toucan was a clever, mischievous bird housed in the zoo's flight cage. Known for his antics, Teeter would steal the zookeepers' keys and hold on to them until the keepers were able to catch up with him and take them back. Teeter had been a resident since the opening of the zoo. (Courtesy of Montgomery Zoo.)

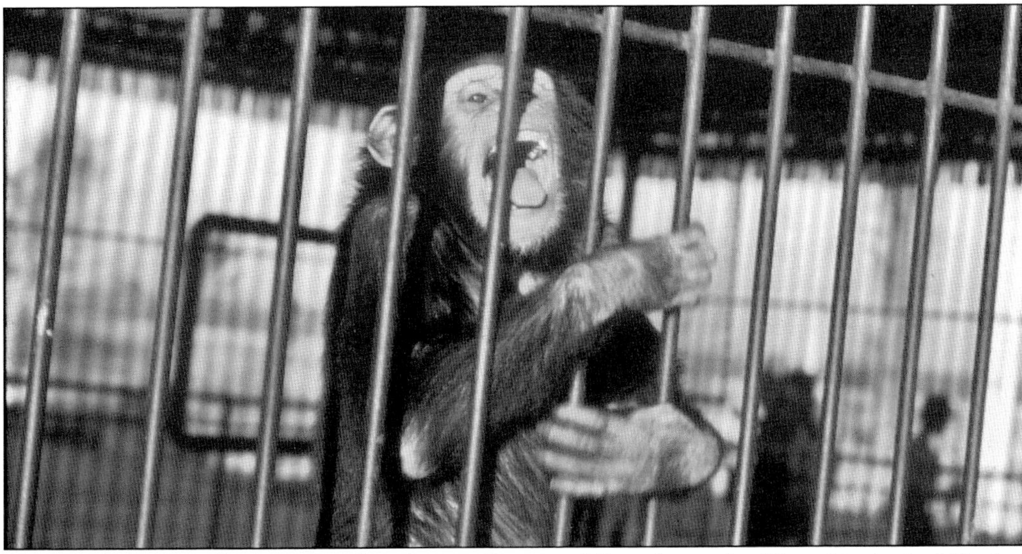

To provide a new playmate for the chimps Benji and Liza Jane, the Dixie Zoological Society donated a two-year-old female chimp named Sonya to the Montgomery Zoo in August 1976. Sonya was raised with a human family in Atlanta before finding her new home with her newly formed chimp family. At first Liza Jane was a little jealous of the attention Sonya received from cage mate Benji, but soon they became friends engaging in many playful activities. To fill her time when visitors were not around and her older cage mates were resting, Sonya would spend her time in other enriching activities such as painting artwork she could then share with her cage mates (below). (Courtesy of Montgomery Zoo.)

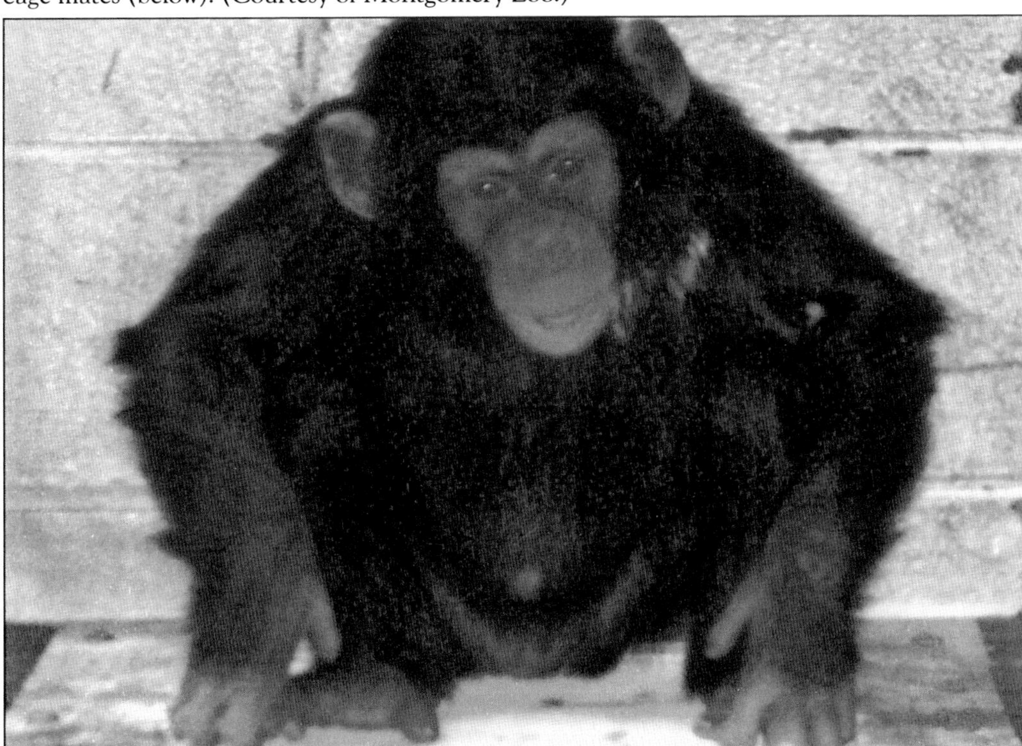

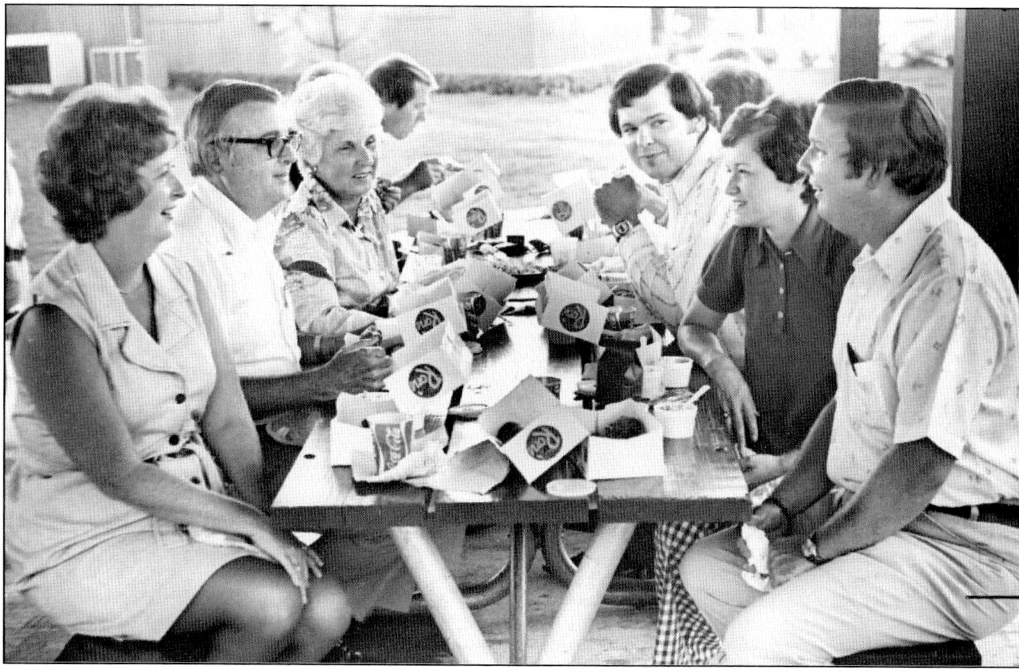

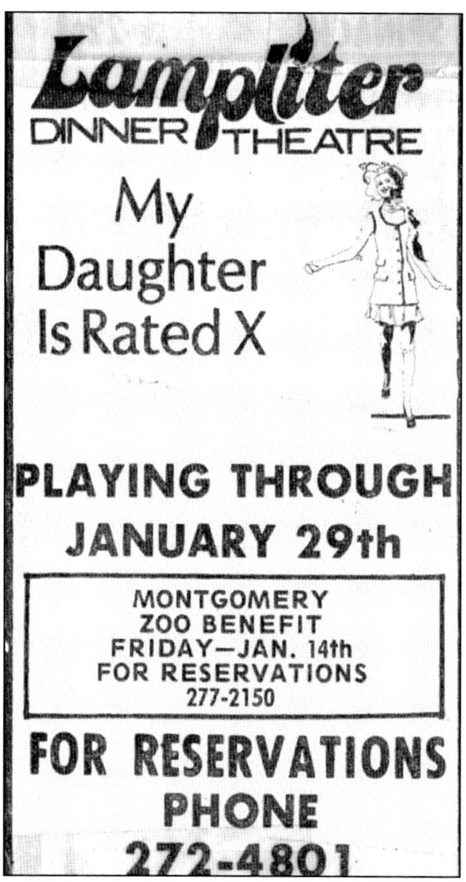

The officers and board of directors of the Dixie Zoological Society held a Saunteering Safari at the Montgomery Zoo on August 11, 1976. The affair began with a box supper, and then zoo director Bird Dog Rogers explained the little-known facts about the zoo inhabitants during an evening zoo tour. (Courtesy of Roger Wentowski.)

In conjunction with a local theater company, the Dixie Zoological Society organized a fund-raising event to obtain the necessary money to purchase two bear cubs and construct a bear exhibit at the Montgomery Zoo. All the proceeds from the January 14, 1976, performance of *My Daughter Is Rated X* at the Lampliter Dinner Theater were donated to the zoo for this purpose. (Courtesy of Montgomery Zoo.)

After only three years, C. B. "Bird Dog" Rogers bid farewell as the Montgomery Zoo director. The zoo was growing quickly, and the park needed a director with the experience to guide the zoo through this growth. In preparation for his departure, Rogers said good-bye to his favorite animals around the Montgomery Zoo. (Courtesy of Montgomery Zoo.)

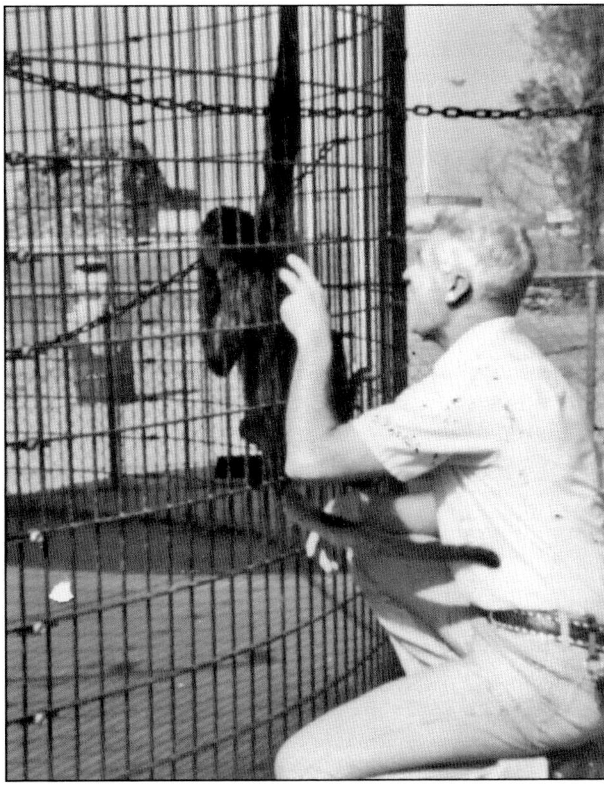

Robert T. Cooper, a graduate of Texas A&M University with a degree in wildlife sciences and postgraduate work in animal physiology and behavior, became the fourth director of the Montgomery Zoo. Before moving to Montgomery, Cooper was the zoo director of Cole Park Zoo in Midland, Texas. With over five years prior experience with zoos Cooper achieved the zoo director title at the young age of 29. (Courtesy of Montgomery Zoo.)

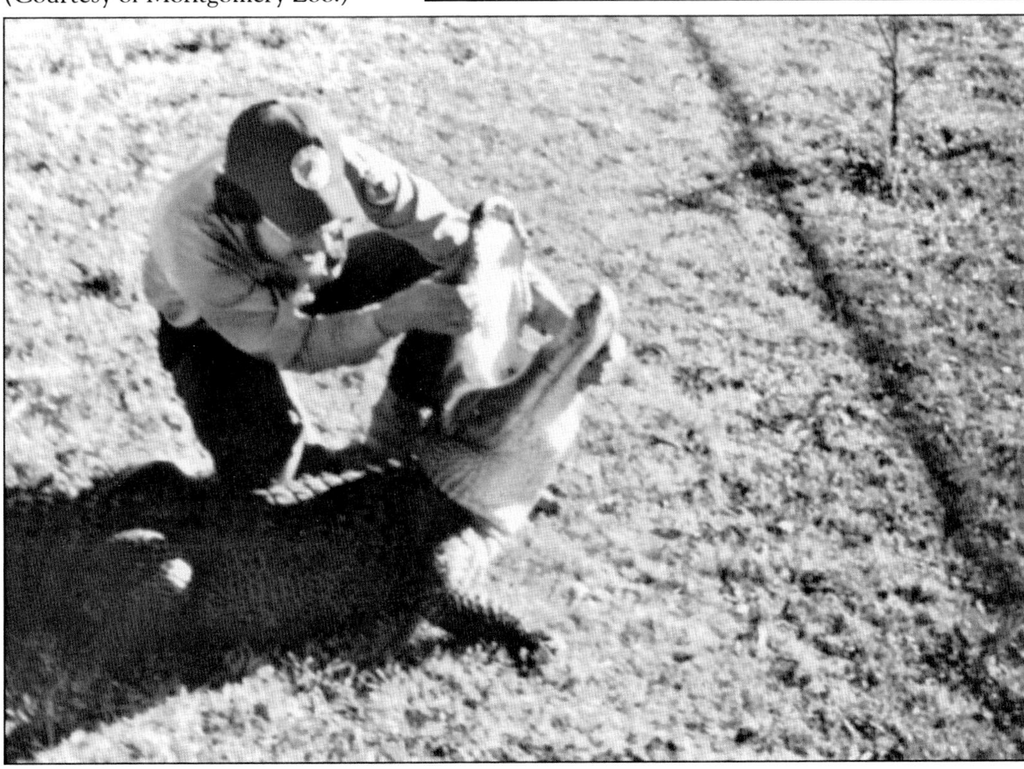

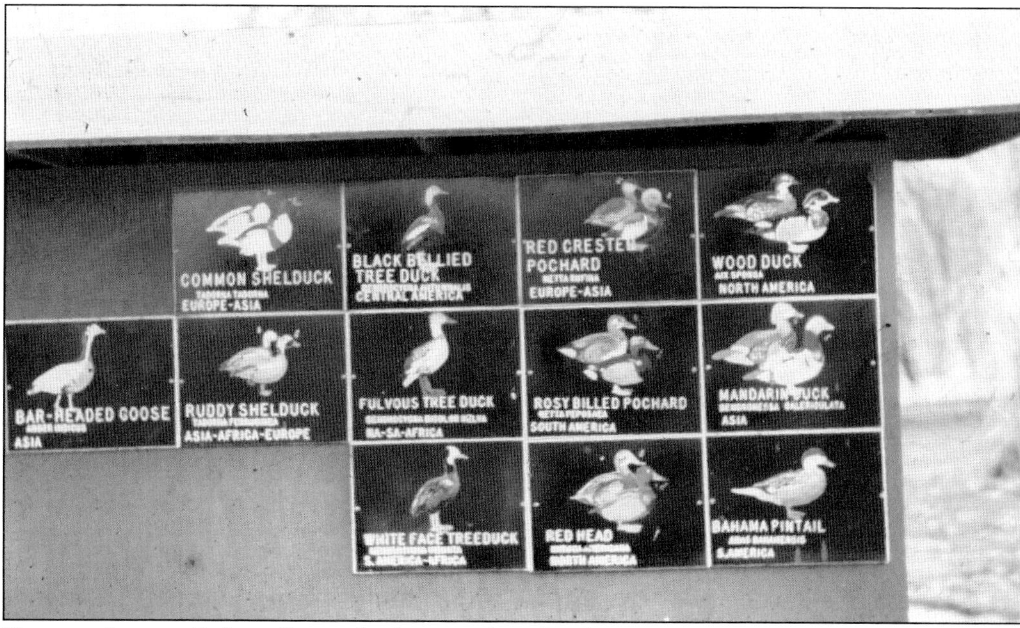

New zoo director Robert Cooper had many plans for the zoo. His main goals included breeding the animals, adding descriptive signs of exhibits, and making enclosures more naturalistic. One of his controversial moves was to get rid of non-breeding animals and replace them with possibly less attractive breeding pairs. (Courtesy of Montgomery Zoo.)

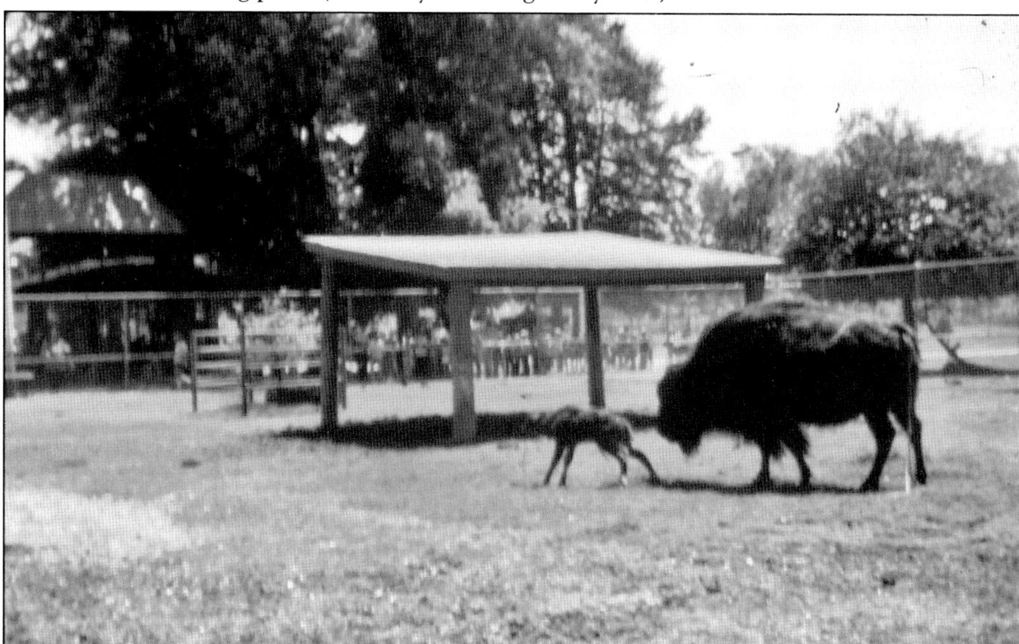

Cooper was more interested in increasing the numbers of animals maintained in zoos through breeding, rather than getting animals from outside sources. To the delight of the new zoo director, the bison of the Montgomery Zoo, Bashful and Boulder, had a baby boy in July 1977. The zoo asked for the public's help in naming the baby and conducted a "Name the Baby Bison" contest. (Courtesy of Montgomery Zoo.)

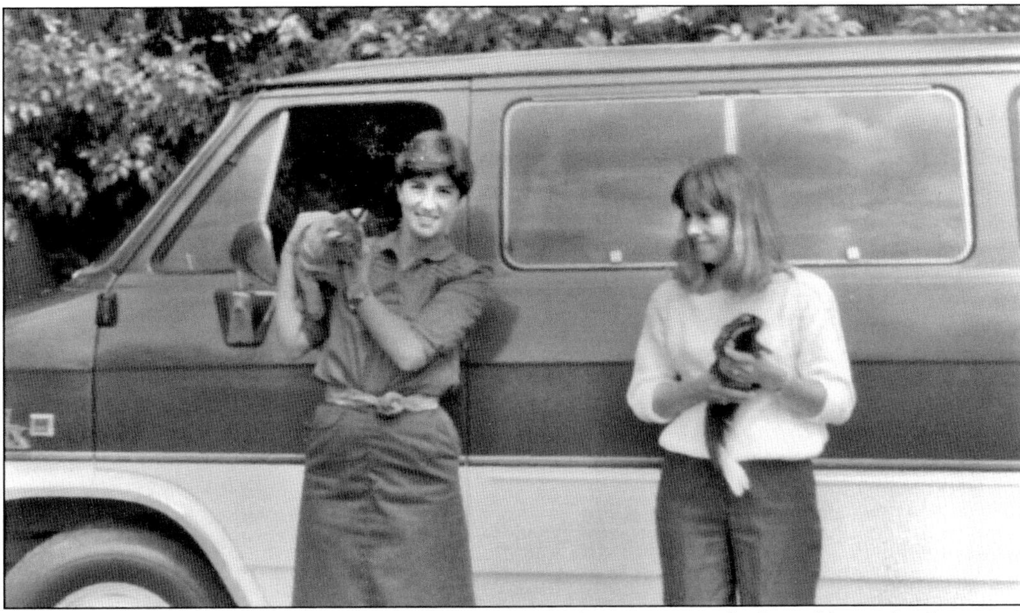

One of the most important aspects of any zoological facility is to provide the public with educational yet entertaining experiences. With the use of volunteers, the Montgomery Zoo was able to develop their educational programs. By taking zoo animals to the public outside the grounds, the zoo can educate people that can not get to the zoo on there own. One such event is seen above with Evelyn Parker (left) and Joan Hodgson, volunteers for the Junior League Zoo Mobile, holding two of the education animals, a bobcat and a frog, as they prepare to leave for their program. Educational animal interactions within the park were also increased to allow keepers to interact with the public while sharing their knowledge of the animals they cared for. (Courtesy of Montgomery Zoo.)

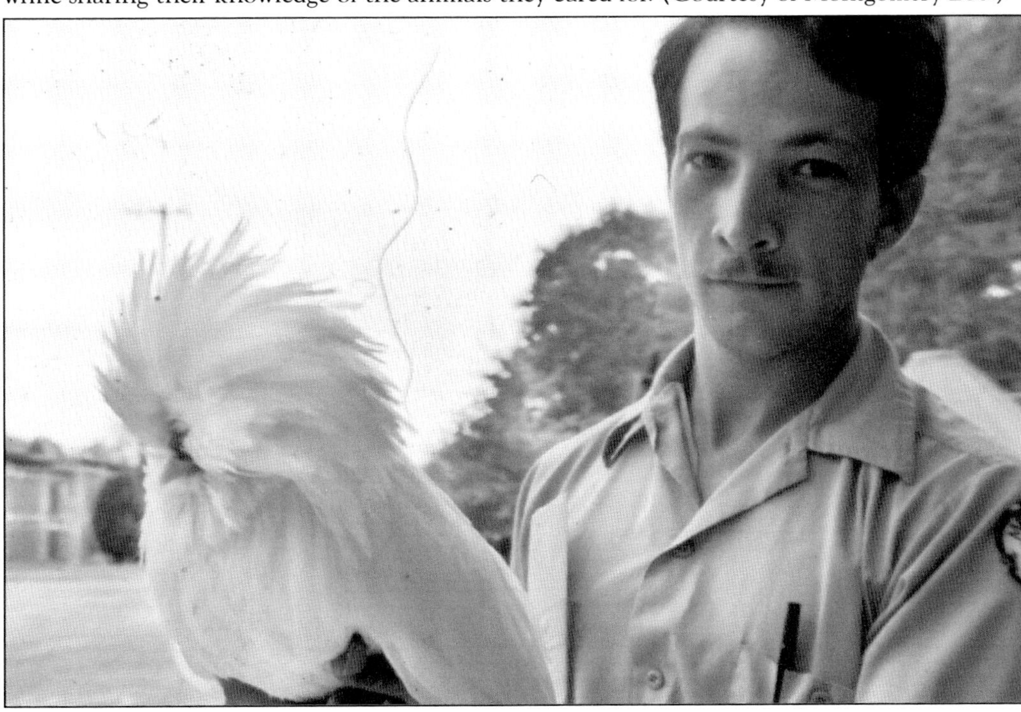

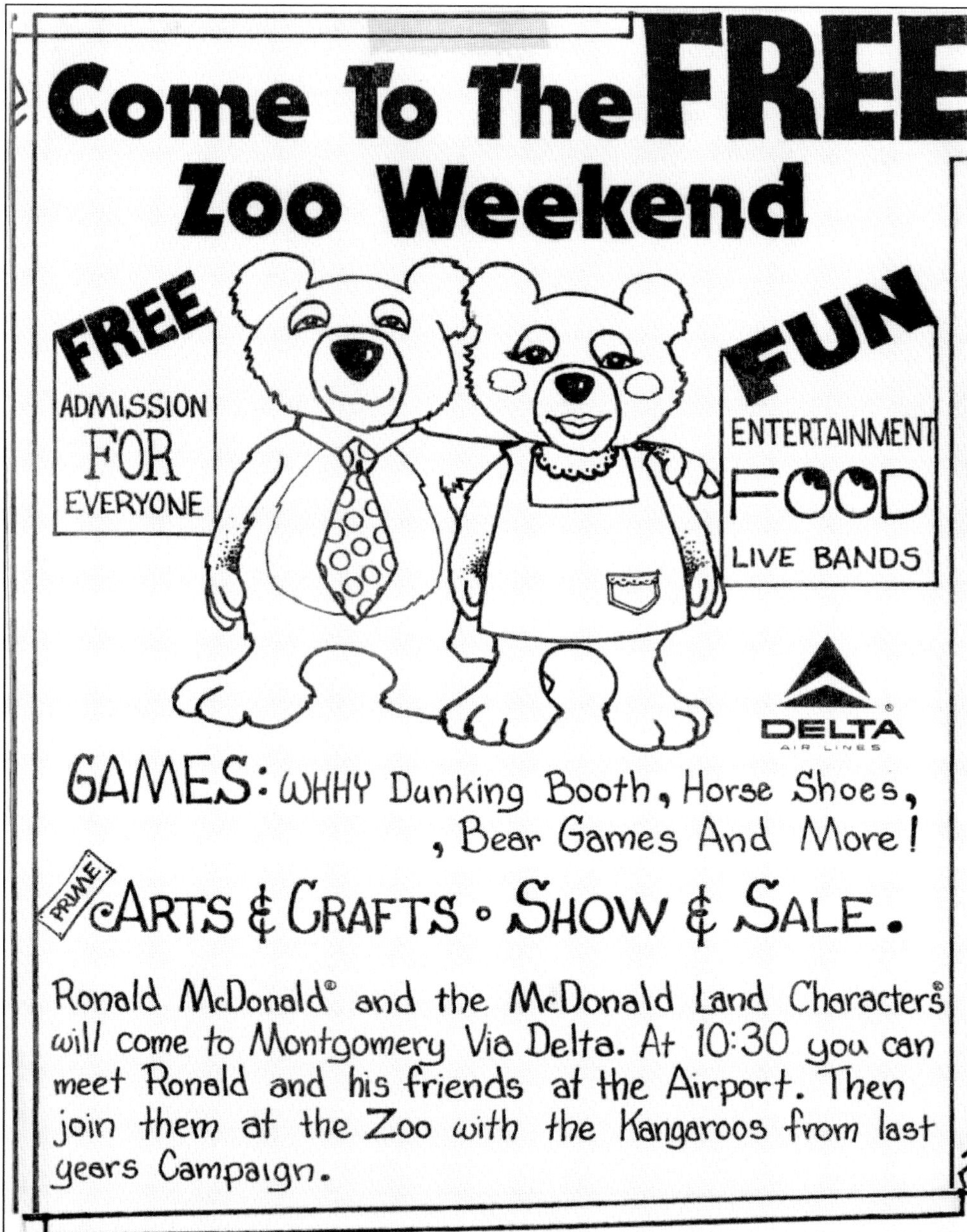

Because of the wonderful success of the first two, the Montgomery Zoo held its third annual Zoo Day fund-raising event in February 1978. This year, the Zoo Day became a Zoo Weekend. To keep this event growing, each year new sponsors needed to be found and asked to come on board. Partnered from the beginning, radio station WHHY continued its support of the event, but this year the Montgomery Zoo also enlisted the support of Delta Air Lines to help create excitement. Part of Delta's participation included having Ronald McDonald and his friends fly into Montgomery from McDonaldland and meet people at the airport. After this successful event, Delta became a yearly participant in Zoo Weekend. Also added to this year's guest list was Snoopy, who walked around the park greeting children and posing for pictures. (Courtesy of Montgomery Zoo.)

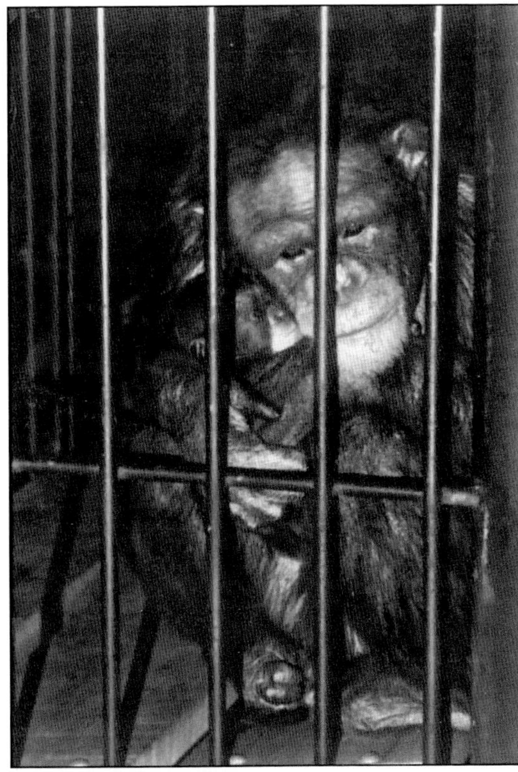

On January 15, 1979, the Montgomery Zoo welcomed their first chimp birth by parents Benji and Liza Jane. The baby boy was quickly named Jimbo. Because both Benji and Liza Jane were hand-raised, the zookeepers were unsure if they would be good parents. No human-reared chimps had ever produced a baby in captivity before, and so the baby's future was uncertain. Keepers watched closely to see how Liza Jane was doing with the baby, and to their amazement, she was a very good mother, although Sonya helped her out a lot with the babysitting. The photograph above shows baby Jimbo being held by his mother, Liza Jane. Below, Jimbo begins venturing out on his own and really puts his teeth into it. (Right, courtesy of Montgomery Zoo; below, courtesy of John Brown.)

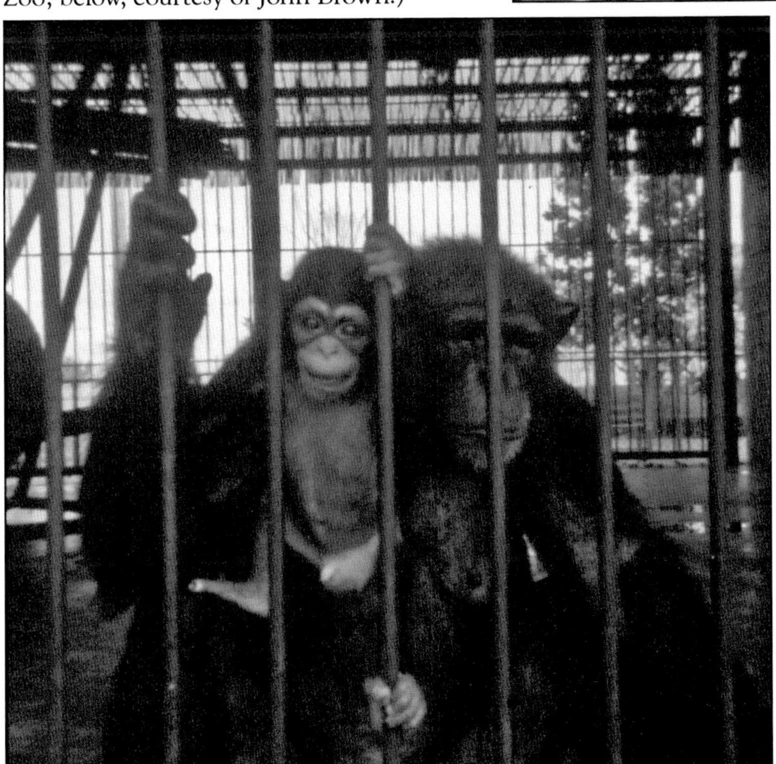

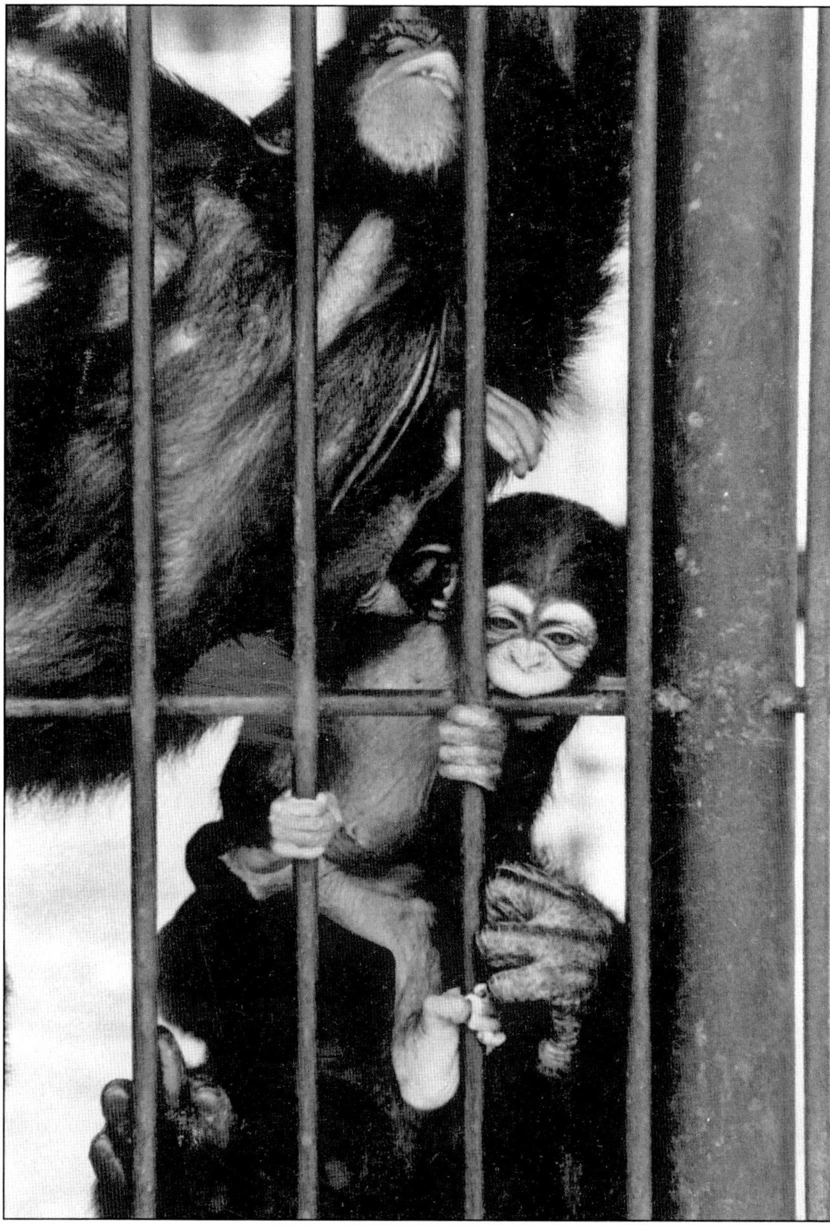

When Jimbo was born, Liza Jane would not allow Benji to see the baby. She kept Jimbo close to her and tried to hide him from Benji's sight. She was being a very smart mother, because when Benji could, he would steal Jimbo from Liza Jane and cause a lot of commotion in the chimp exhibit, with both Liza Jane and Sonya chasing after him trying to retrieve Jimbo. Keepers would come running to monitor the situation and make sure that this important baby was returned unharmed to its mother. Soon the keepers realized that these "kidnappings" were occurring only when no one was at the chimp exhibit and were soon settled after their arrival. Realizing that Benji was taking Jimbo to get attention, they would no longer come to the exhibit when this would happen but would watch the situation from a distance, unknown to Benji. When the attention stopped, so did the kidnappings, leaving the females to raise the baby, as is proper in chimp society. (Courtesy of Montgomery Zoo.)

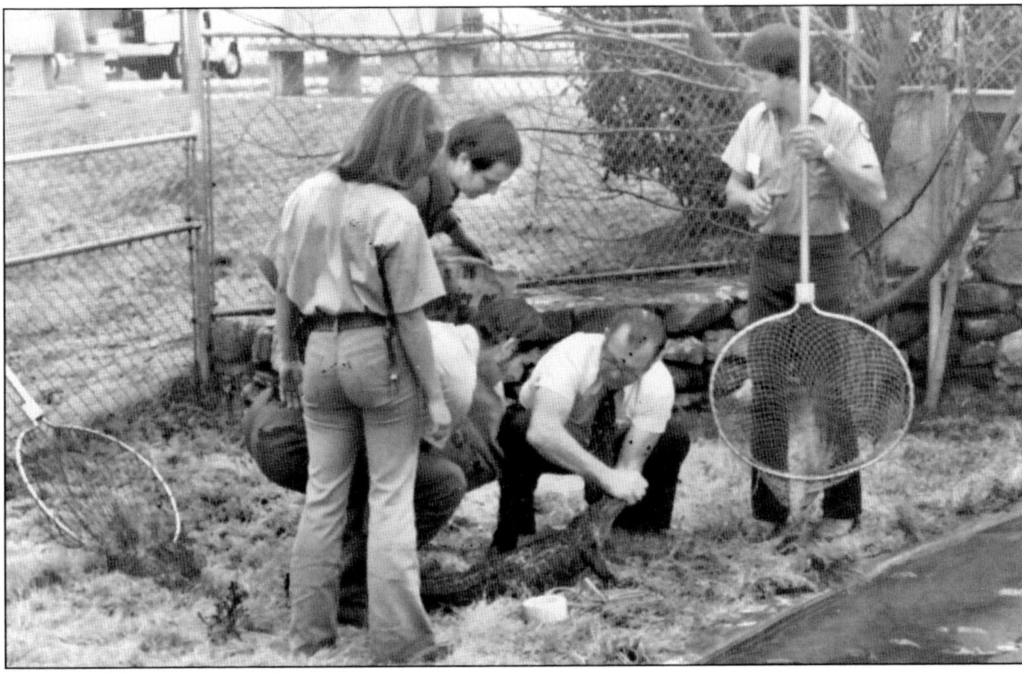

Chuck Clift replaced Bob Cooper as zoo director of the Montgomery Zoo on January 31, 1979. A native New Yorker, Clift was director of the Syracuse Zoo for seven years, Pittsburgh Zoo for two years, and Warner Brothers Jungle Habitat in New Jersey for three years. Clift gets down and dirty helping the keepers wrangle an alligator. (Courtesy of Montgomery Zoo.)

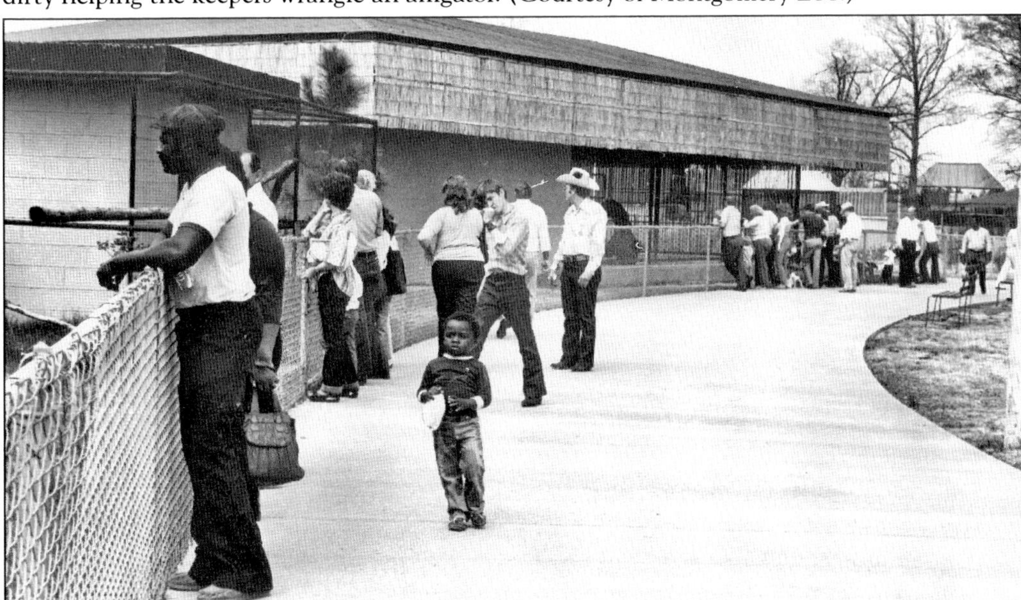

New zoo director Clift had a vision of creating the fastest-growing zoo in the entire Southeast. By creating mix-species exhibits and more natural interactions of animals, he would be able create a more appealing zoo for both animals and visitors. Mixed-species exhibits, at the time, were mostly seen with birds and petting zoos. Clift was planning to apply this innovative approach with larger animals in naturalistic habitats. (Courtesy of Montgomery Zoo.)

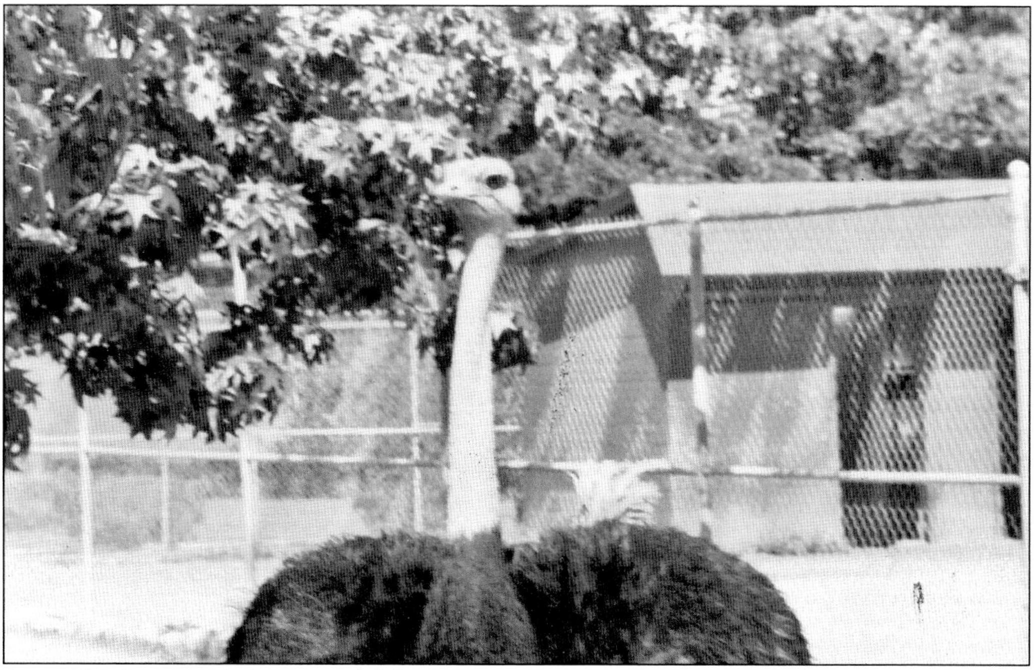

Soon after his arrival, director Clift brought new animals into the park. Augmenting the ratite collection, two friendly ostriches (above) were added. These two outgoing birds could be seen interacting with their keepers, pulling hair and following them around their exhibit. Later that year, he obtained South American Patagonian cavies to be displayed in a mix-species exhibit with the South American rhea, starting the current trend of exhibiting animals in geographical realms. Patagonian cavies are large rodents closely related to the guinea pig, an animal the zoo often used as an educational tool with visitors. With this new addition, zoo educators could use the cavies in an interesting way, allowing the guests to interact with the guinea pigs while learning about their wild relatives. (Courtesy of Montgomery Zoo.)

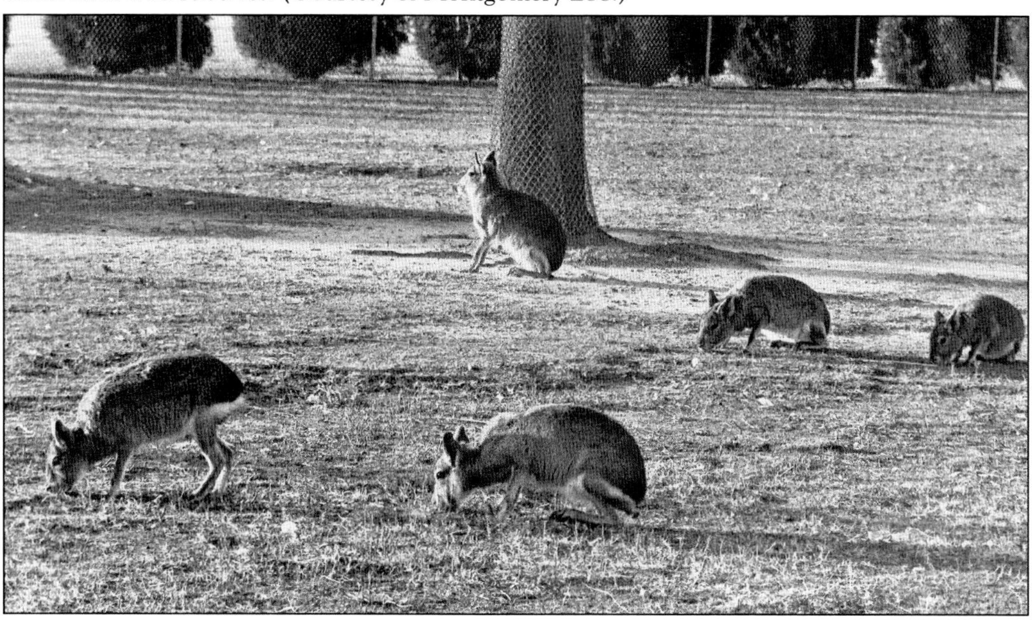

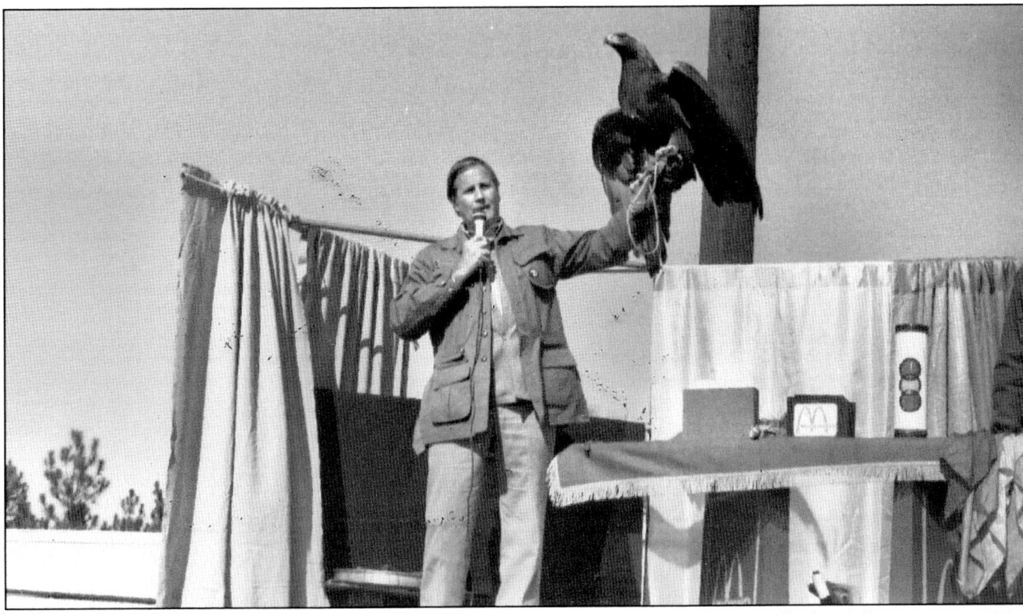

To continue the tradition of Zoo Weekend and entice visitors to attend, the Montgomery Zoo invited another named personality as their featured guest. Jim Fowler, known for his works on television's *Mutual of Omaha's Wild Kingdom*, interacted with the visitors and public officials during Zoo Weekend 1979. Other attractions during the weekend included Ronald McDonald, arts and crafts, and games for the children. Sponsors continued to increase for the Montgomery Zoo's Zoo Weekend to included radio station WHHY, Ronald McDonald, the Montgomery Zoological Society, and Delta. A special welcoming committee headed by Montgomery mayor Emory Folmar was on hand to welcome Ronald McDonald and featured guest Jim Fowler. Zoo Weekend 1979 was an even greater success than had been hoped for, and the zoo successfully raised the money needed to upgrade the lion exhibit. (Above, courtesy of Montgomery Zoo; below, courtesy of *Delta Digest*.)

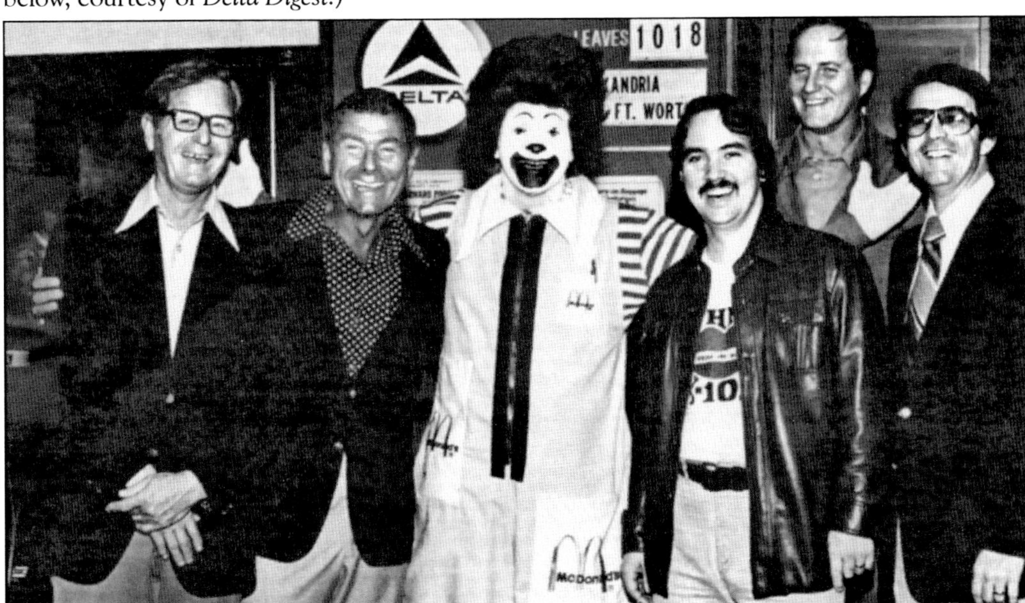

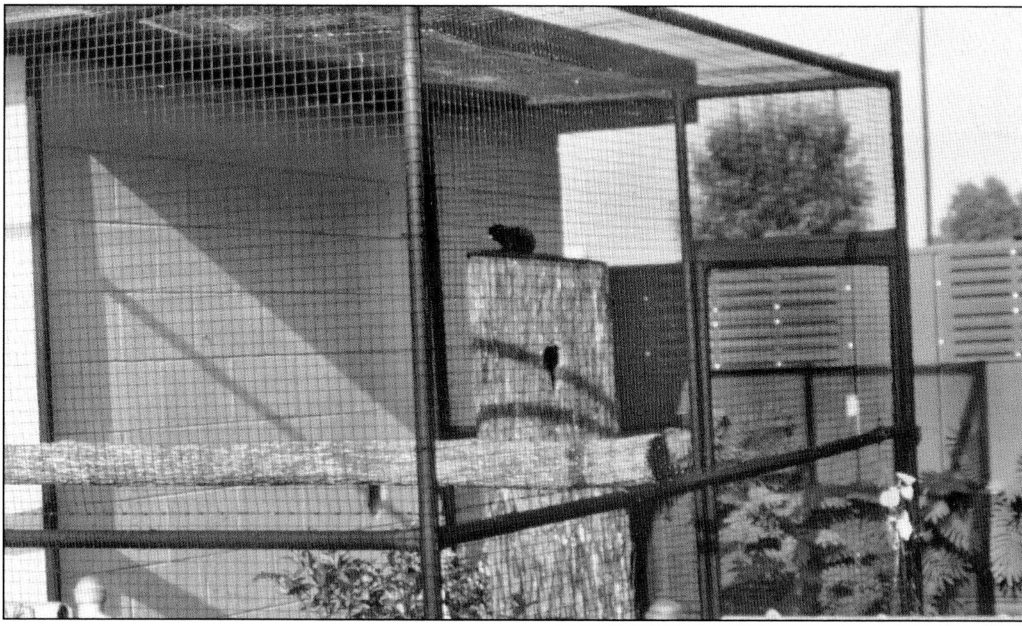

On the morning of July 15, 1979, when the keepers arrived at the zoo, they were greeted by two unexpected arrivals. While conducting morning rounds of the animals, they found in the brown-headed tamarin exhibit, riding on their fathers' back, two baby brown-headed tamarins. This was the first tamarin birth at the zoo. (Courtesy of Montgomery Zoo.)

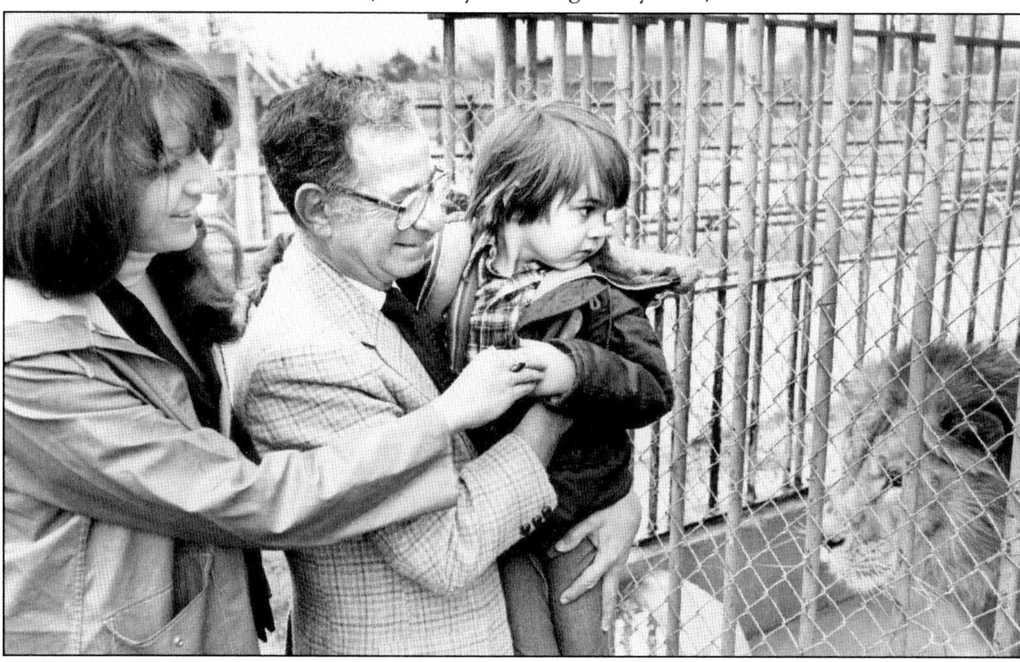

Barbara Keel, an artist, created a portrait of Myrobi and Big Mamma (before she passed away) to be entered in to the annual art contest at the zoo. She then surprised the zoo by donating half the proceeds of the original painting and giving 300 signed copies to the Montgomery Zoo for them to sell. The money was donated to fund a new lion exhibit for Myrobi. (Courtesy of Montgomery Zoo.)

The animals went back to Oak Park when Festival in the Park was held there in September 1979. The Montgomery Zoo brought animals to create a display for the festival. The zoo brought a monkey and macaws that people could look at and set up a petting-zoo area where they could touch and even feed goats, sheep, and deer. Bambi, a white-tailed deer fawn, was a highlight for the children. With no admission charge, thousands of people flocked back to Oak Park to visit the animals. Animals were not the only attraction, though. Festival in the Park also had arts and crafts and games for the kids. (Courtesy of Montgomery Zoo.)

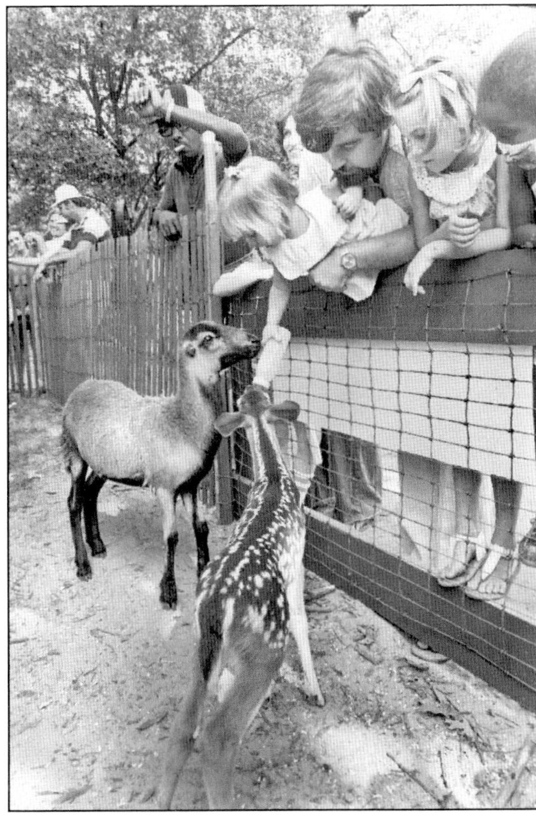

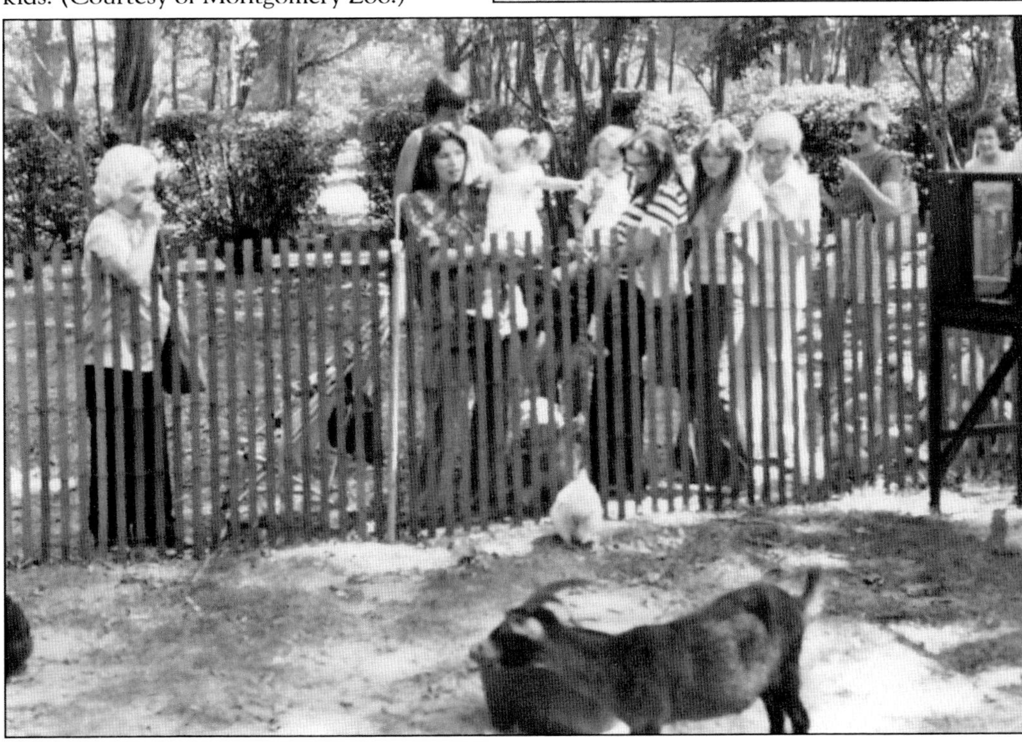

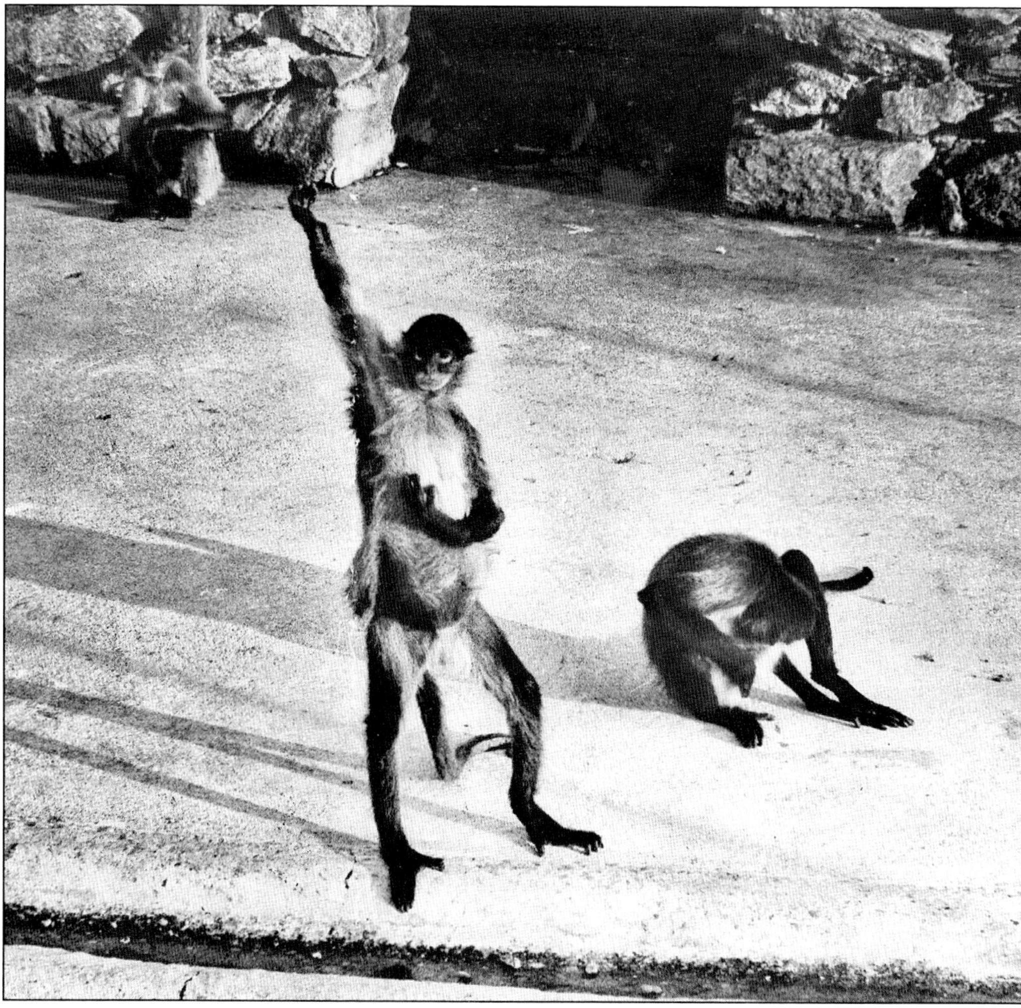

Otis, a male capuchin, was living with three male spider monkeys on Monkey Island. Before coming to the Montgomery Zoo, Otis lived at the Fort Worth Zoo, where he had previously escaped his habitat. At the end of 1979, Otis escaped from Monkey Island at the Montgomery Zoo for the first time. He was caught by zookeepers using a baited cage because he kept outrunning them. As a former pet, keepers believed that Otis was being bullied by the much larger spider monkeys. Then a few months later, he escaped a second time and visited children and a pet cat in a nearby neighborhood. After being recaptured for the second time, *That's Incredible* wanted to film Otis, but he suddenly died of a virus in 1980. Before his death, a female capuchin was put with Otis in his habitat off the island. It is believed that Otis somehow told Cocoa how to escape Monkey Island, because she was the only other monkey to be successful at escaping. (Courtesy of Sheila Bergquist.)

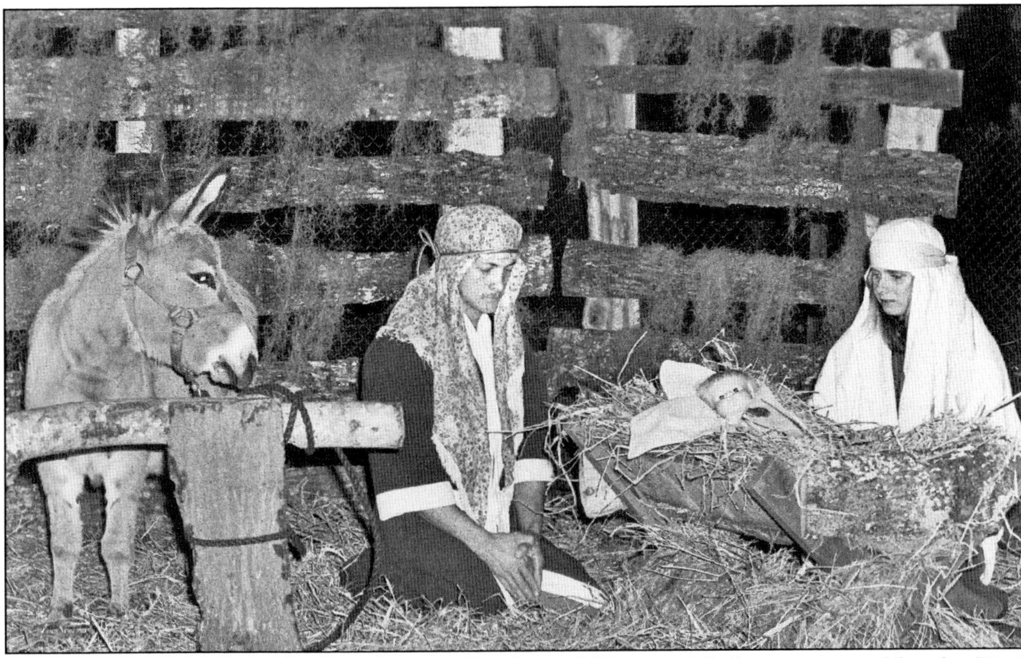

On December 22, 1979, the Montgomery Zoo hosted Nativity Nights, sponsored by the Dixie Zoological Society. This annual events exhibited animals from the zoo: two camels, two donkeys, three goats, and three sheep. Nativity Nights included a musical pageant using animals and people to create the scene at the manger. Zookeepers dressed in costumes played the shepherds, so the animals were handled safely during the event. The Music Crusader choir of the First United Methodist Church and the Montgomery Boys Choir provided musical performances, while WHHY's Larry Stevens narrated the event on the radio. This popular evening became an annual event at the zoo, growing in size each year. (Courtesy of Montgomery Zoo.)

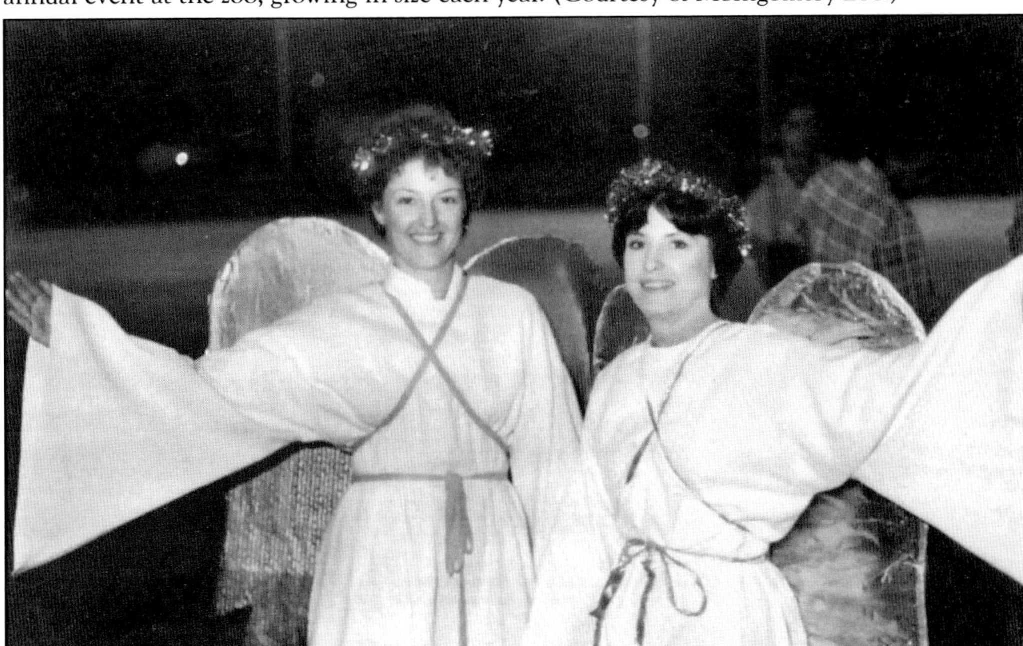

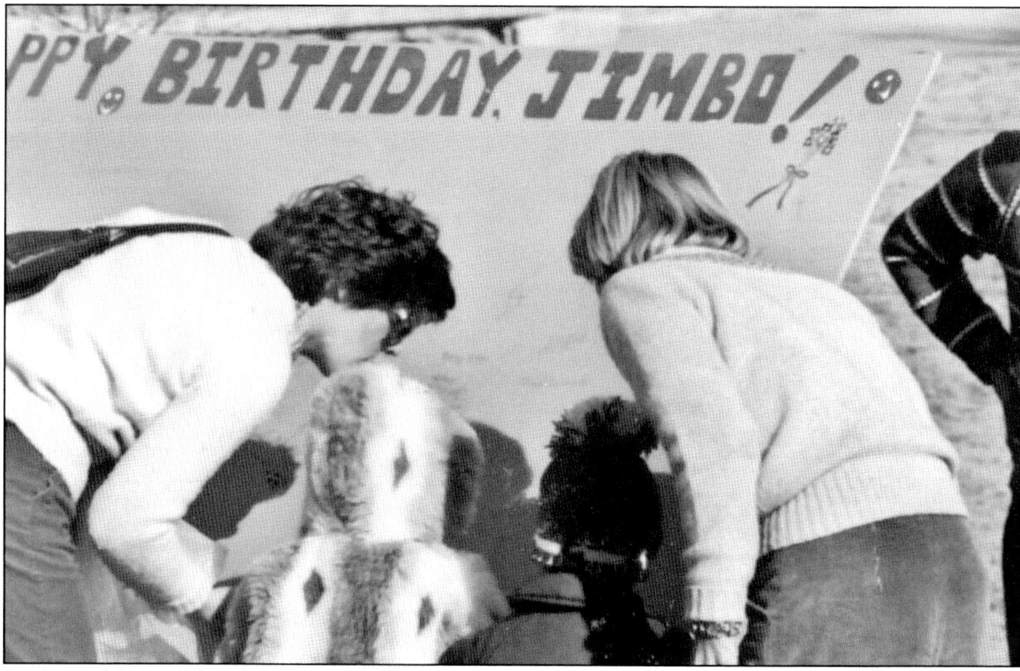

Jimbo celebrated his first birthday on January 16, 1980. In order to celebrate the incredible achievement of Liza Jane and Benji in raising their young chimp, a large party was planned. As part of the day's festivities, a large birthday card (above) was available for visitors to sign. A special birthday cake (below) was prepared for Jimbo, made with fruit, carrots, and monkey chow. Another more traditional birthday cake was shared with his many human guests after they all sang him "Happy Birthday." Admission was free for visitors that brought a present or a camera. (Courtesy of Montgomery Zoo.)

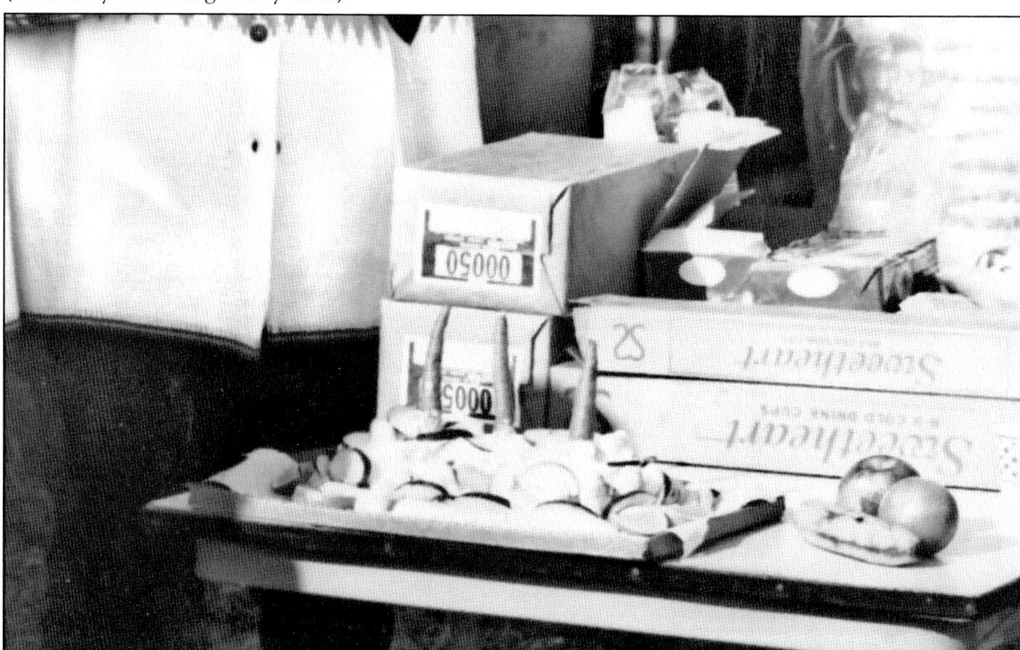

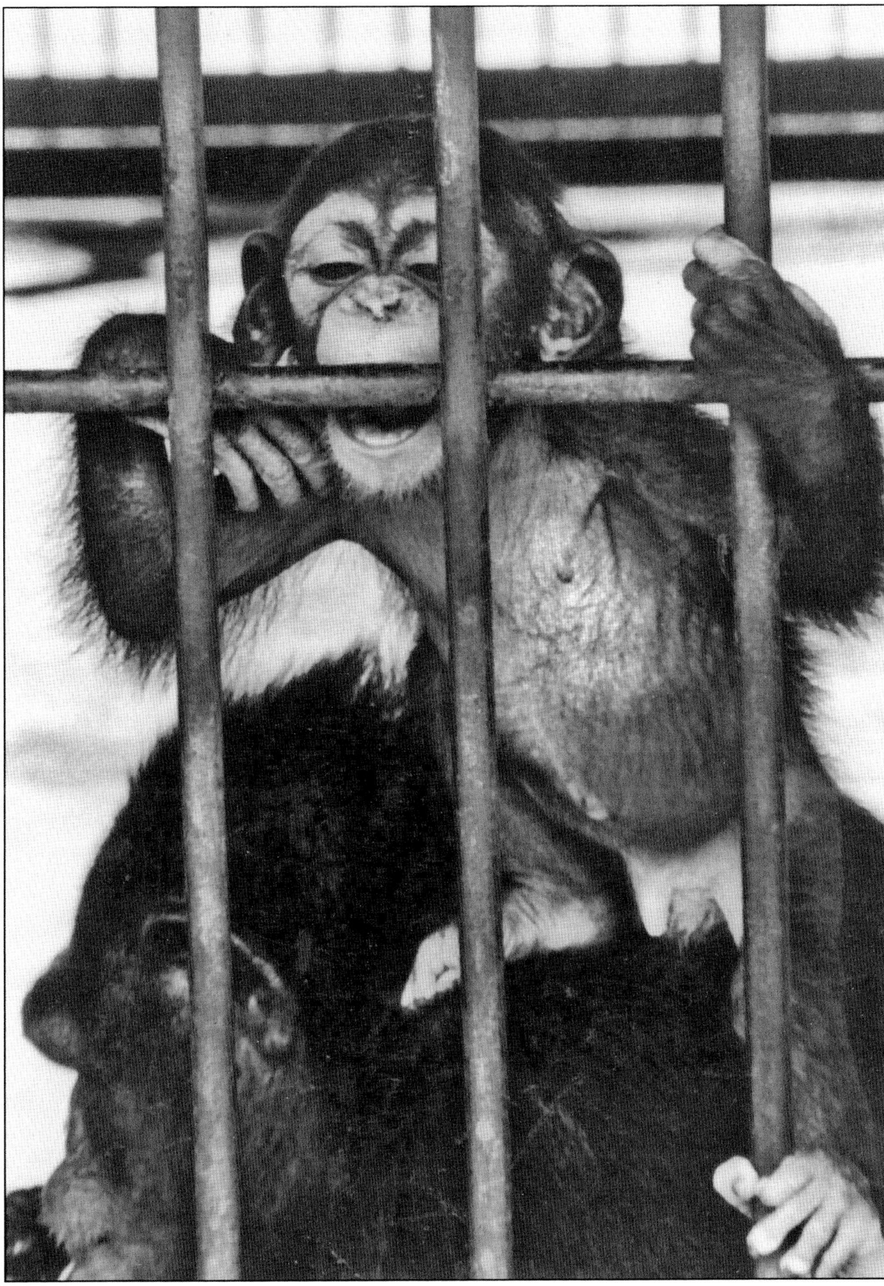

To prepare for the party, zoo volunteers arrived early at the chimp enclosure. After wishing their special boy a happy first birthday, the work began. The exhibit was decorated on the outside with balloons for the guests to enjoy and colored paper inside for the chimps to play with. Cake and drinks were prepared for the guests that were expected to attend, and tables for people to leave their presents. Though one-year-old Jimbo was living with his parents, Benji and Liza Jane, and aunt, Sonya, he seems to know that all the attention was for him. In acknowledgement of the attention he received from the large crowds formed for his birthday, Jimbo ventured away from his mother and out on to the bars of the enclosure, allowing everyone a good look at him as he played. (Courtesy of Montgomery Zoo.)

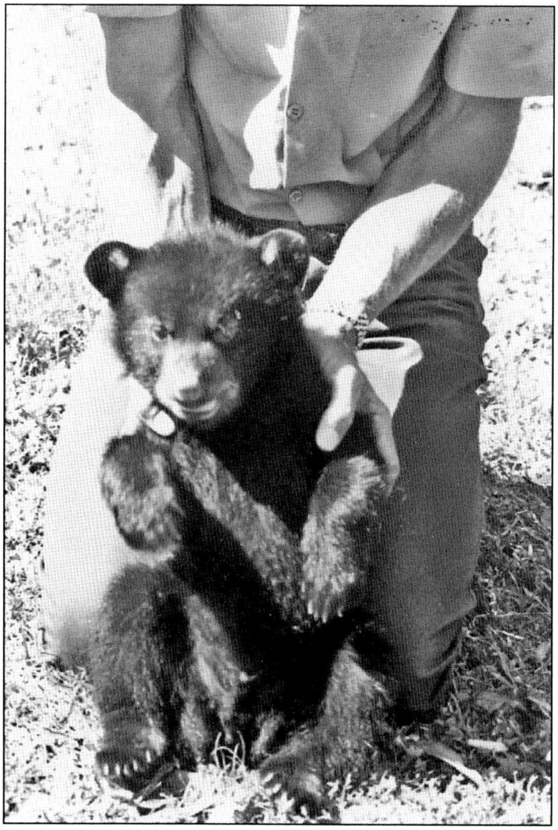

After four years of fund-raising, the Montgomery Zoo received two female North American black bear cubs from Knoxville, Tennessee, in June 1980. Unfortunately the exhibit was not completed, so they were placed in the spot-nosed monkey cage. A "Name the Baby Bear" contest was held, and Ke'sha and Ne'sha were selected. (Courtesy of Marsha T. Lautenbach.)

In November, the bears were finally introduced to their new exhibit, which was dedicated in the memory of Florian Strassburger, a chapter member of the Dixie Zoological Society. Canuk, a little male black bear from the Toronto Zoo, joined the girls shortly afterward. He was an orphan. His mother was shot in Northern Ontario, and her three little cubs survived by being brought to the Toronto Zoo. (Courtesy of Montgomery Zoo.)

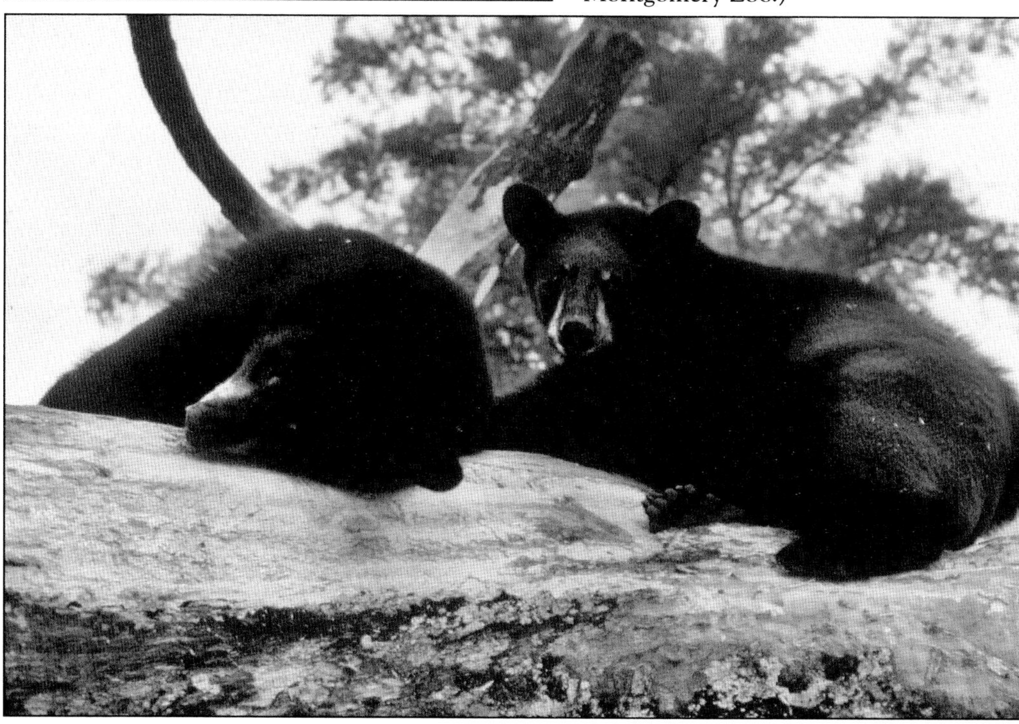

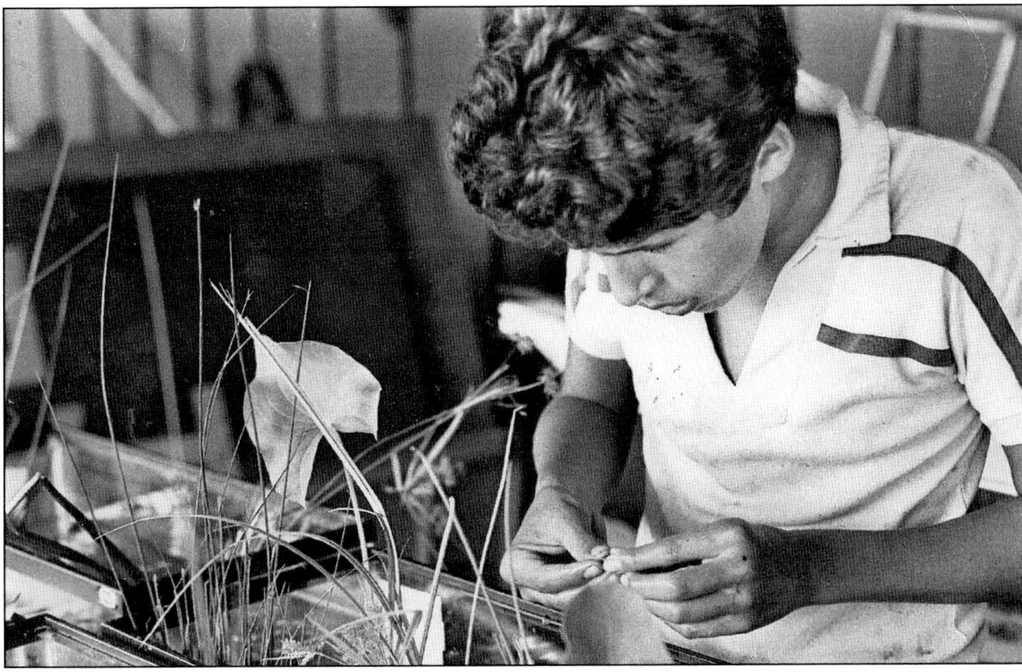

The Montgomery Zoo began an educational summer program called the "Discovery Program" for kids 13 to 16. Charlie Pullinger, age 14, checked out a mollusk before putting it into an ecosystem he was creating. This was a two-week wetland ecology experience for students to develop a keen awareness of the diverse life of the wetlands. The Discovery Program became a yearly program that provided an escape to the great outdoors for 25 teenagers who wanted to get back to nature. Students also studied microscopic life forms and learned to identify inhabitants of area forests and waterways. (Courtesy of Montgomery Zoo.)

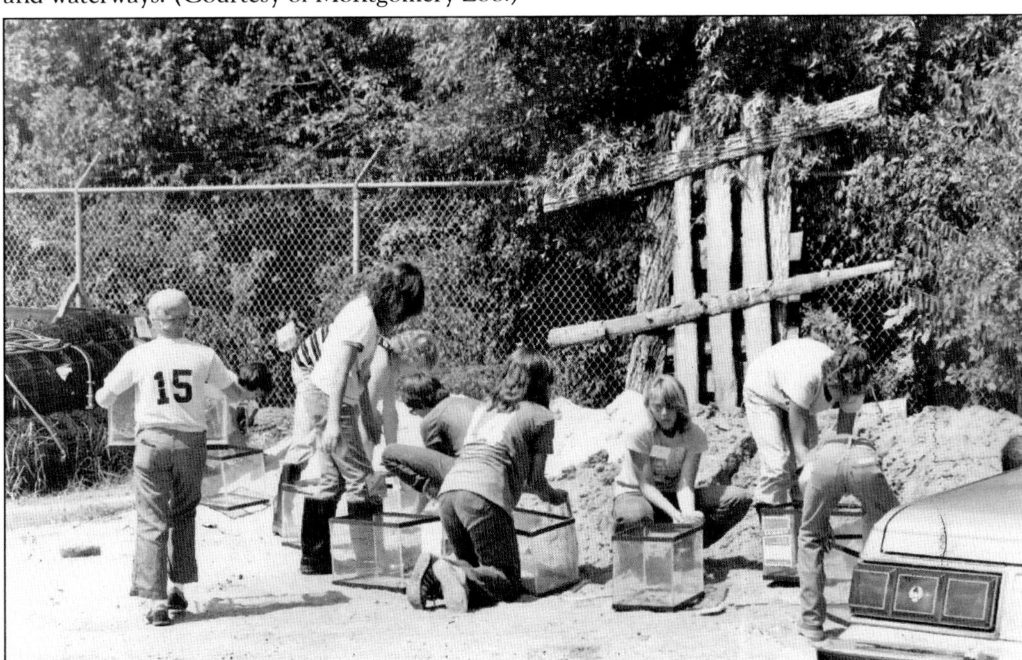

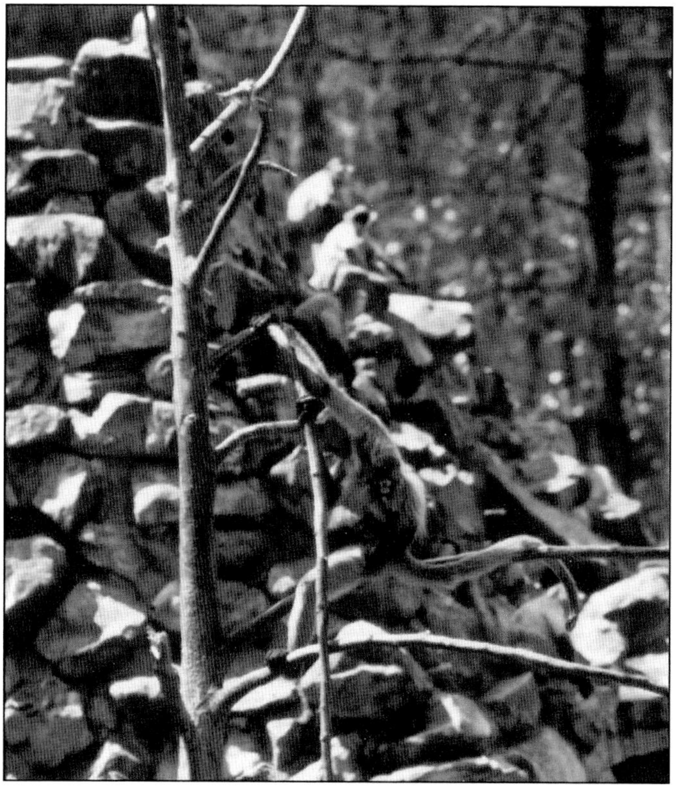

The zoo was happy to report their first spider monkey baby. One of Chuck Clift's missions for the zoo was to have successful breeding pairs. As one of the most popular animals at the Montgomery Zoo, the spider monkey group was growing and continued to expand in the coming years. (Courtesy of Montgomery Zoo.)

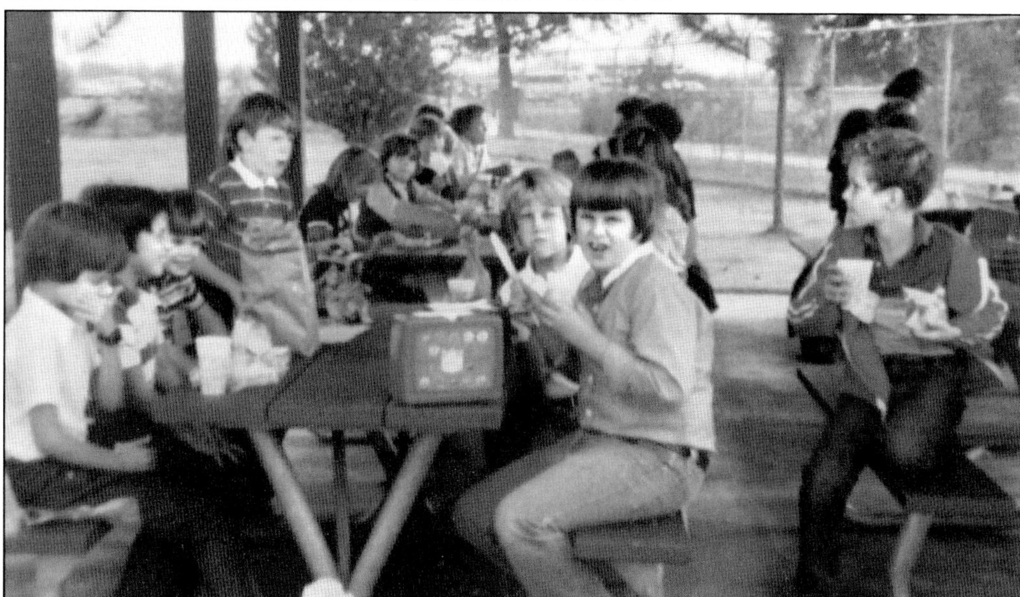

The Annual Enrichment Program took place in the fall of each year at the zoo. Through this program, children developed an understanding and appreciation of animals and their roles in life. The Annual Enrichment Program was one of several programs designed to meet the needs and interests of the public. The fifth-grade class of Mrs. Faulkner and Miss McDowell visited the zoo to participate in the program. (Courtesy of Betty Breslin.)

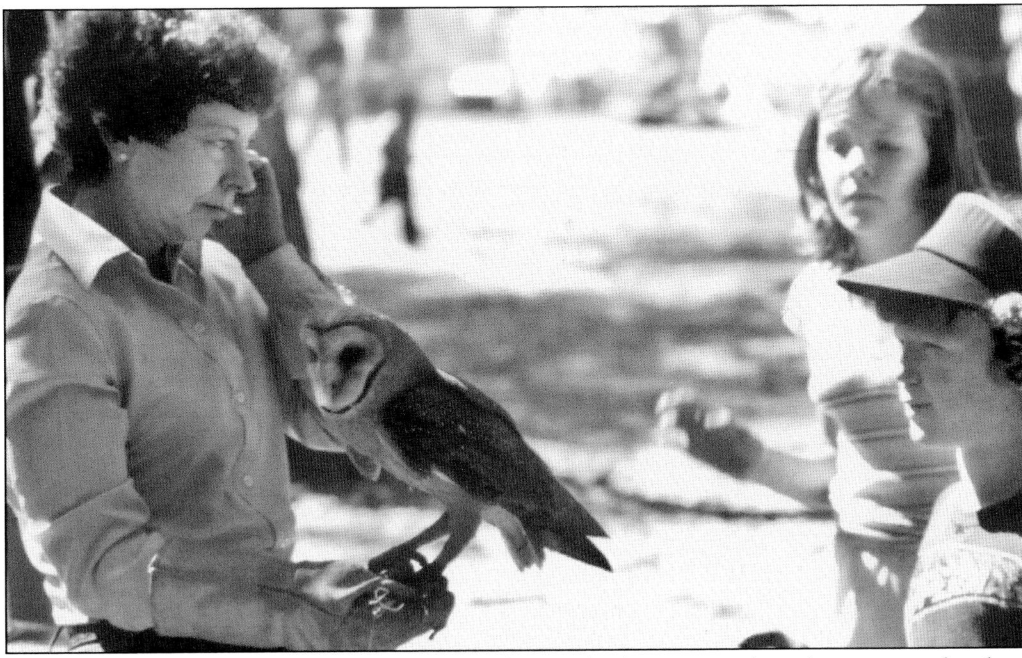

ZOOREACH, the Montgomery Zoo's program that reached out to people with messages of ecology, conservation, and wildlife appreciation, received the Alabama Environmental Award of Excellence in Education in 1981. Taking the educational aspect currently being used by zookeepers and docents at the zoo, ZOOREACH was created as a more formal program. This new educational approach brought the animals out into Montgomery, allowing for a more interactive educational opportunity. One of ZOOREACH's many programs included visiting nursing homes with animals that could be held. Volunteers also took animals to local schools and libraries to provide children with fun educational facts. (Courtesy of Montgomery Zoo.)

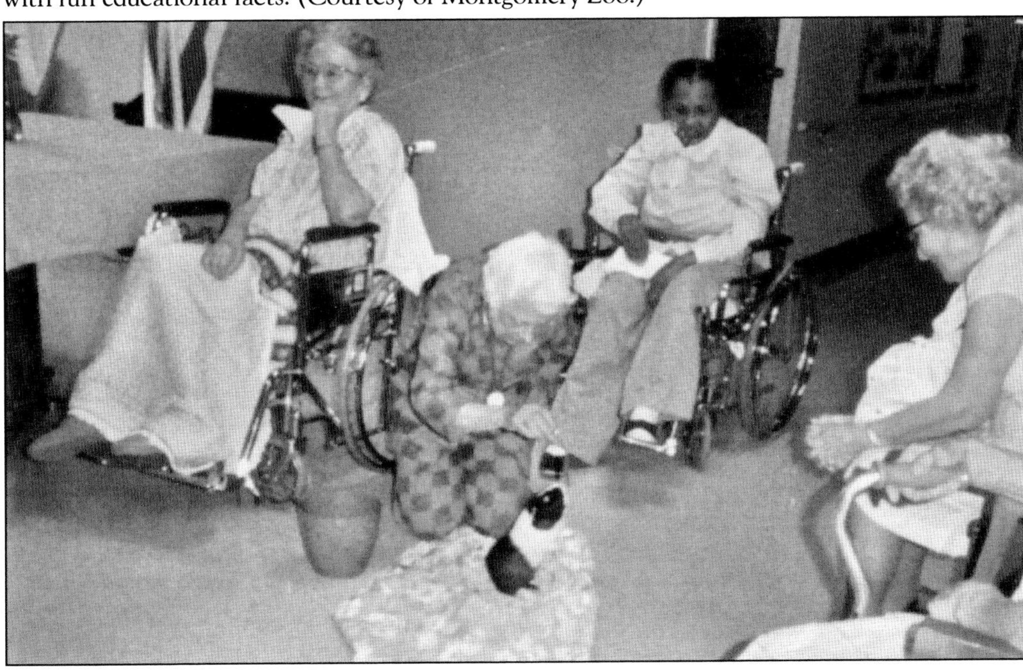

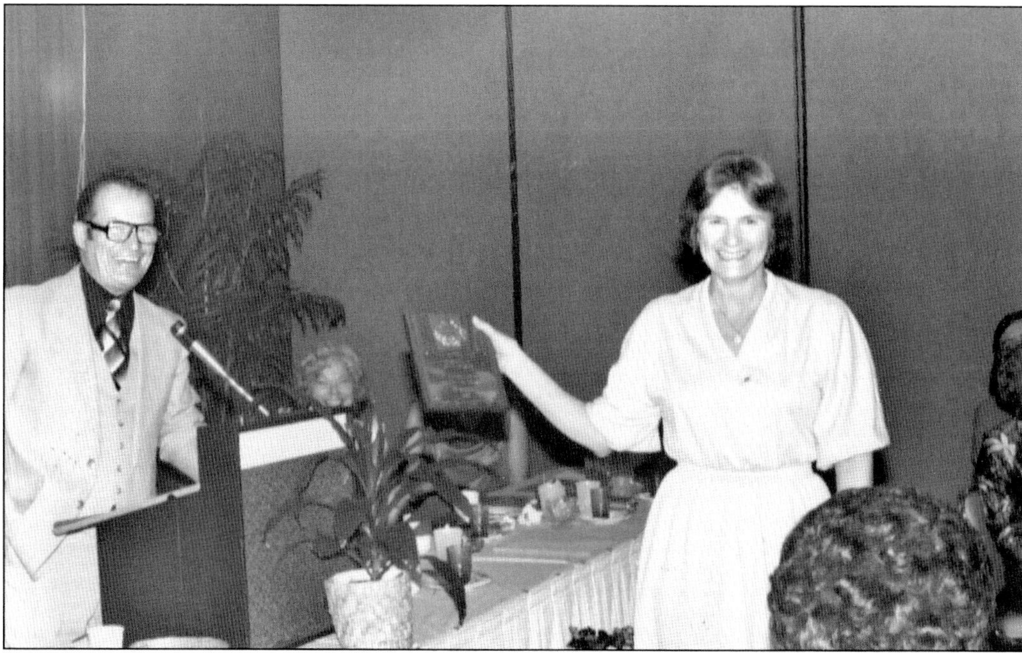

Betty Breslin, a long-time volunteer at the Montgomery Zoo, received an award from VIZA (Volunteers in Zoo Activities) for her work at the zoo. The annual awards banquet occurred on January 31, 1981. Chuck Clift presented Breslin with her plaque. (Courtesy of Betty Breslin.)

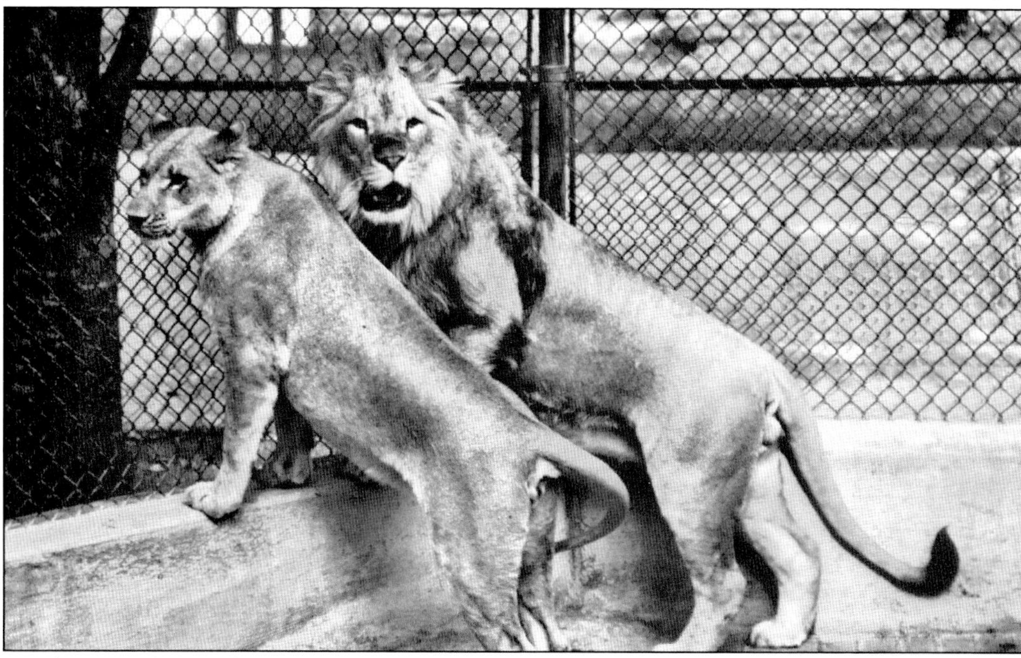

Fund-raising during the seventh annual Zoo Weekend, held in March 1982, went toward the new lion exhibit, already in construction. At the same time, the zoo received a female lion from the University of North Alabama. Suzie, a six-year-old lion, was previously a house pet that was given to North Alabama. However, they were unable to introduce her to their lion and found her a new home in Montgomery. (Courtesy of Montgomery Zoo.)

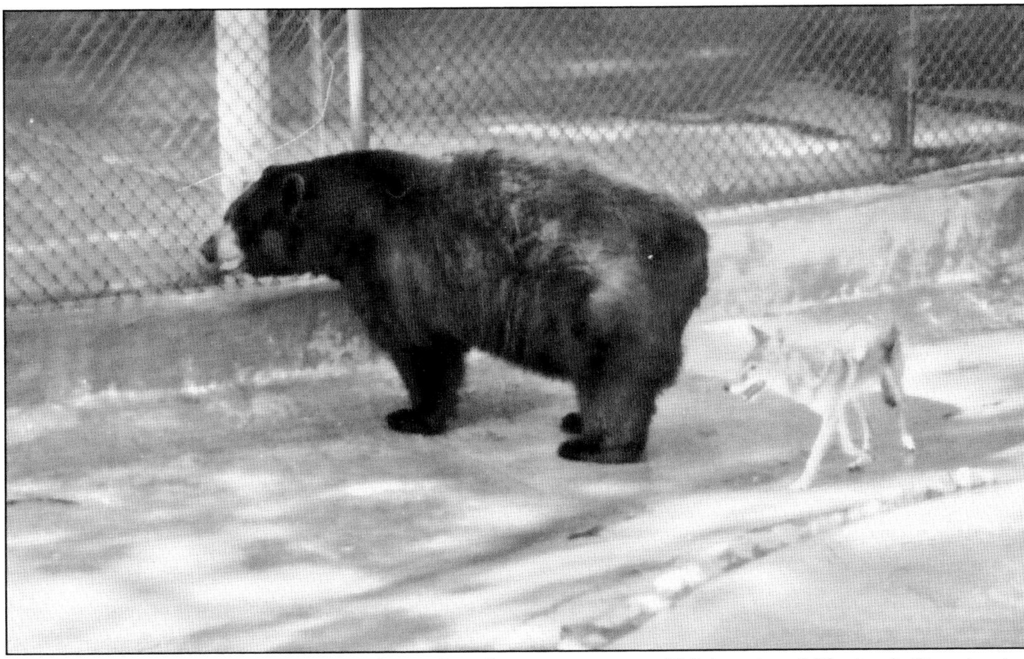

In the fall of 1982, the zoo acquired two female coyotes named Nahani and Hasinai. Continuing with his vision of mix-species exhibits, Clift introduced the two coyotes into the black bear exhibit, creating an "Endangered Species in Alabama" exhibit. It was a very successful exhibit that eventually yielded four bear cubs. (Courtesy of Montgomery Zoo.)

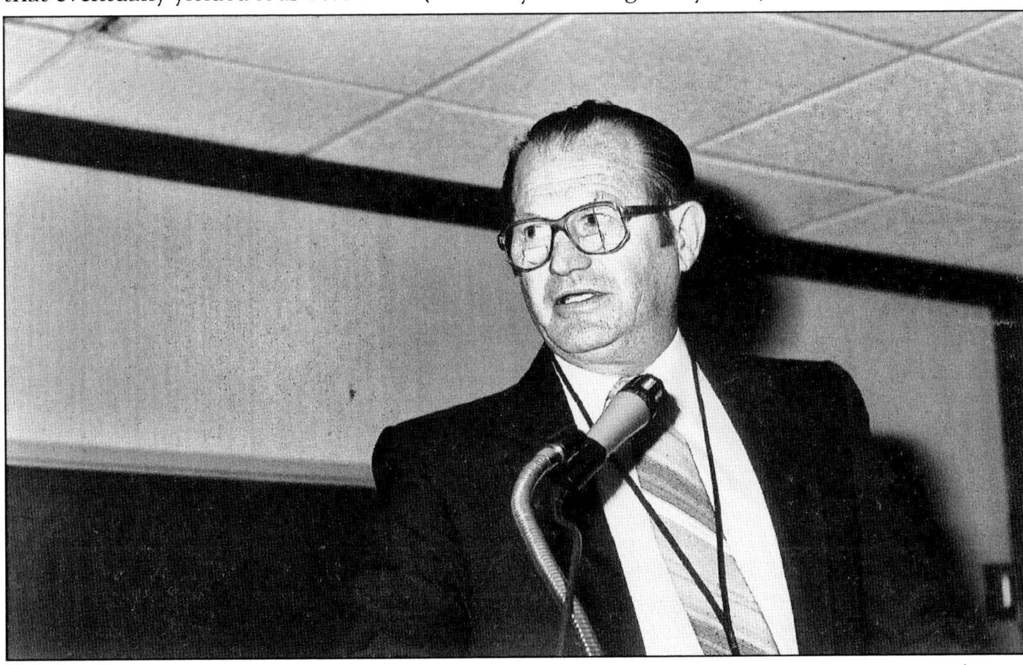

Zoo director Chuck Clift lost his eyesight during surgery to remove a brain tumor near the end of 1981. Mayor Folmar asked Clift to remain on as the director of the Montgomery Zoo. To get around better, Clift memorized the location of every animal in the park allowing him to still be productive. (Courtesy of Montgomery Zoo.)

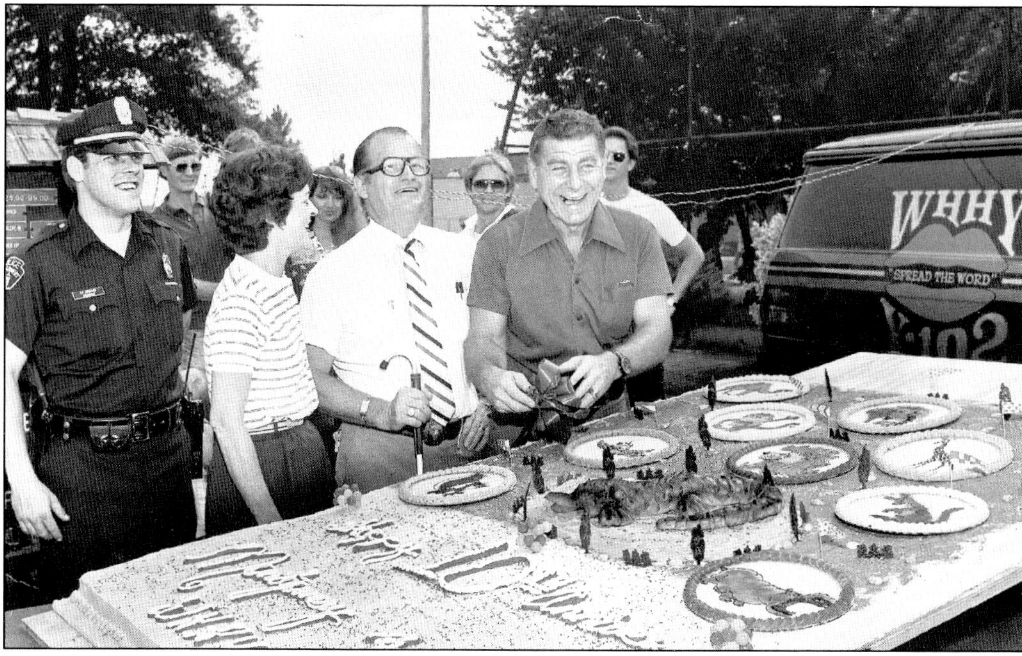

The Montgomery Zoo celebrated its 10th birthday on August 7, 1982. During the two-day celebration of the zoo's maturity and growth, many special events were planned, including the dedication of the new lion/cougar exhibit and entertainment by musical groups, magic acts, square dancing, and performances from the Montgomery Little Theater. A large birthday cake was also shared with all the guests. Shown cutting the first slice from left to right are Anita Folmar, zoo director Chuck Clift, and Mayor Emory Folmar (above). The new cat exhibits (below) were funded completely through money raised during previous zoo weekends and thus were truly a present to the zoo from the people of Montgomery through their support of such events. (Courtesy of Montgomery Zoo.)

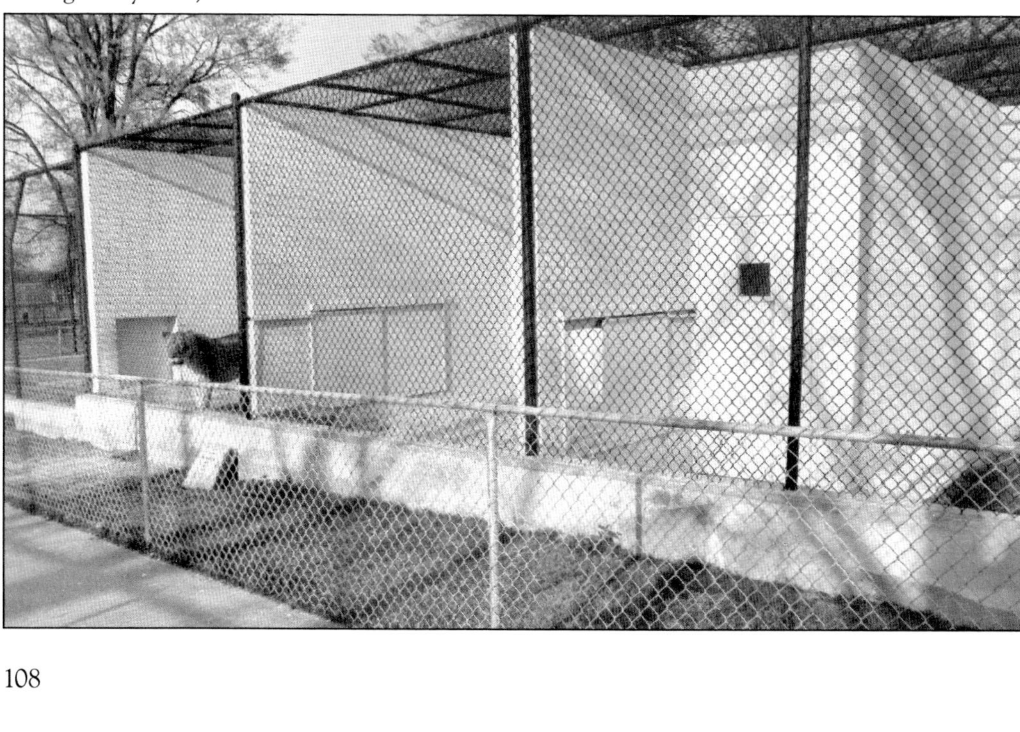

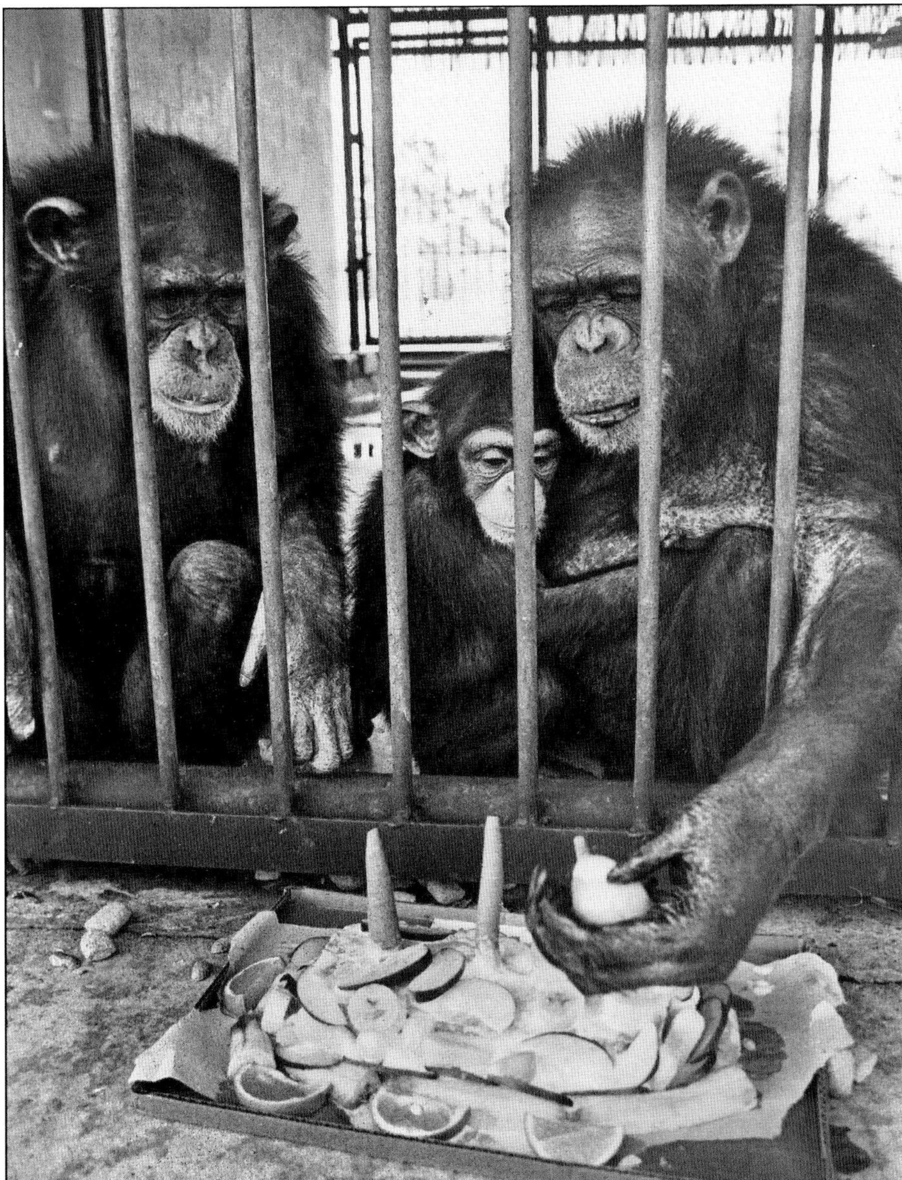

Jimbo, the zoo's most famous resident chimpanzee, celebrated his third birthday in January 1983. Like in previous years, a gala party was prepared for Jimbo, his family, and all his friends from the Montgomery area. A special cake topped with vegetables was presented to the chimpanzees for their enjoyment. There were several exciting activities during his special day, including a coloring contest sponsored by McDonald's. Though the chimpanzees enjoyed the day, this would be one of the last birthdays they would spend together as a family. A couple of years later, the Montgomery Zoo would bid farewell to their beloved Jimbo. In order to uphold the zoo's commitment to conservation, Jimbo was sent to the Cleveland Zoo in Ohio on a breeding loan; breeding male chimpanzees were rare in the zoo world at that time, and the Cleveland zoo had four females eagerly awaiting Jimbo's arrival. While growing up at the zoo, Jimbo had learned several entertaining tricks, like standing on his head, begging, and blowing kisses, that were missed by both his keepers and the public. (Courtesy of Montgomery Zoo.)

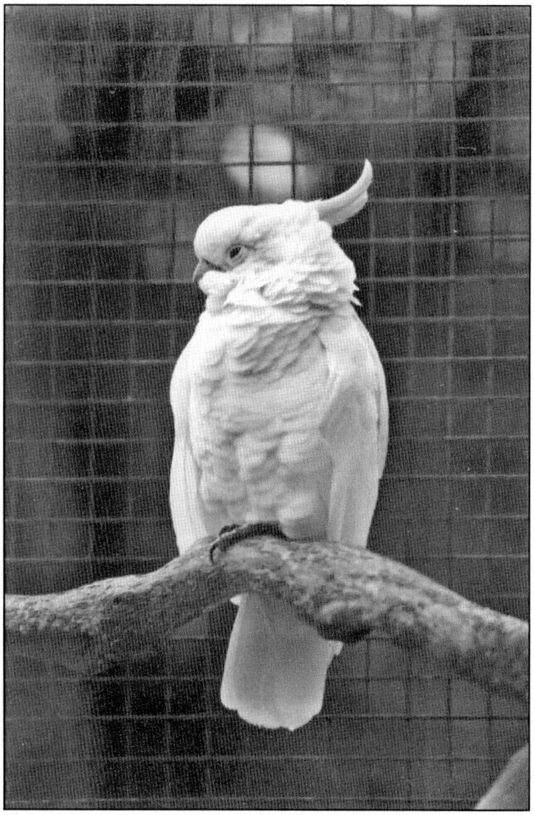

The Montgomery Zoo held a male citron crested cockatoo, which is a highly endangered bird. The Central Alabama Avicultural Society donated a female citron that could breed with the male. These were the only two citrons in the state of Alabama. The female bird was shipped from Indonesia so that the Montgomery Zoo would have the opportunity to breed one of the rarest species of cockatoo around. (Courtesy of Montgomery Zoo.)

The Eighth Annual Zoo Weekend of 1983 featured Joan Embrey, animal trainer and goodwill ambassador for the San Diego Zoological Society. The new bear and lion exhibits were open for the public to visit. Known for her appearances on the *Tonight Show*, Embrey tried to save animals from becoming extinct. (Courtesy of Montgomery Zoo.)

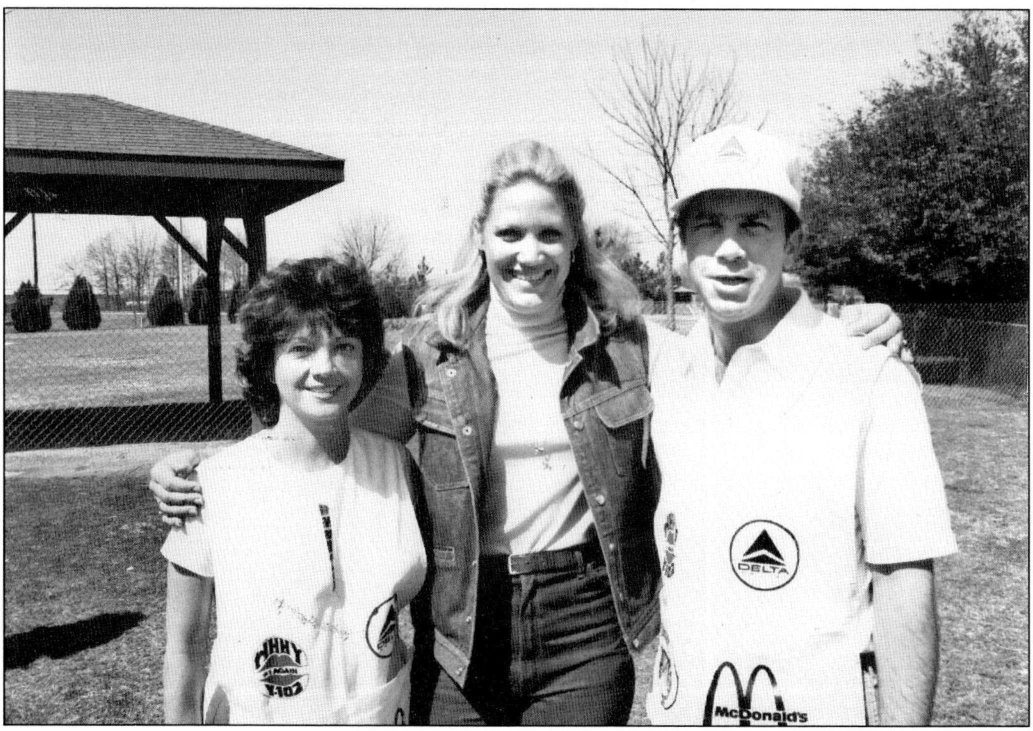

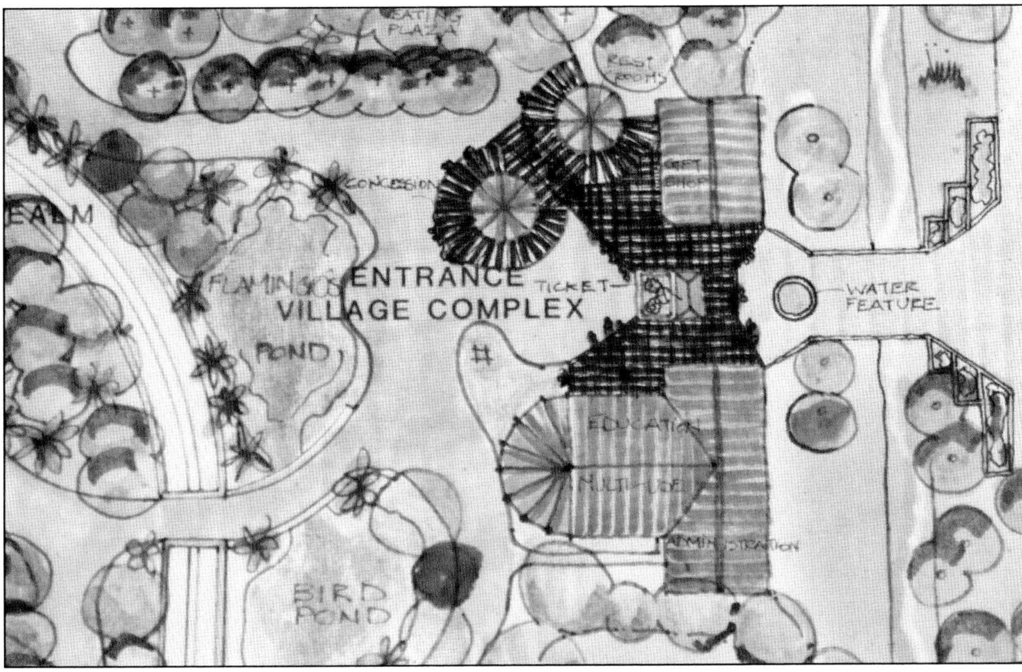

Thirty-four acres of land was made available to the Montgomery Zoo in 1983. The zoo at the time occupied six acres of land. The new acreage was to be used for future expansion as monies became available. The Dixie Zoological Society began working on a long-range master plan for the zoo. McFadzean and Everly, Ltd., were selected as the contractors for the expansion. (Courtesy of Montgomery Zoo.)

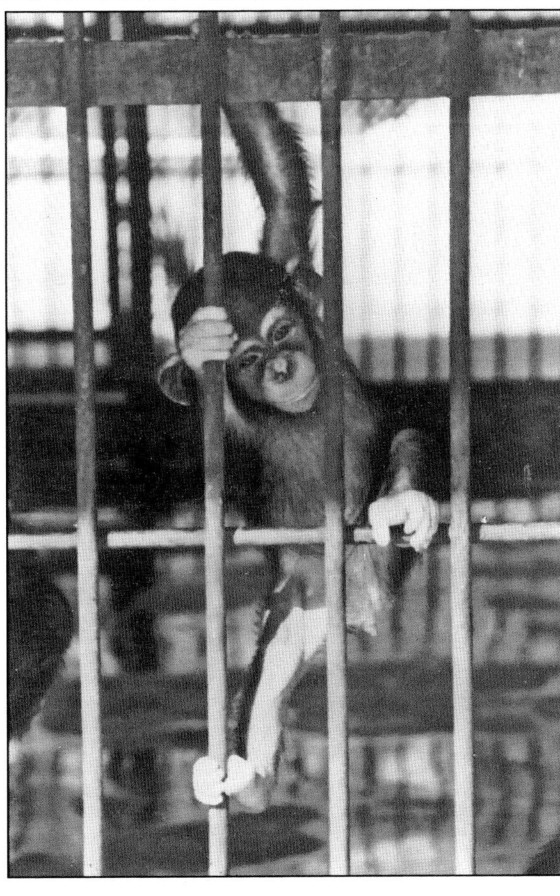

The Montgomery Zoo welcomed another baby chimp. Miss T was born on April 3, 1986, to Benji and Sonya. Like Benji, the zoo celebrated her birthday with large annual events. She also penned a column where should would talk about other residents of the zoo. Named in honor of city council member Billy Turner's wife, Miss T, became an instant celebrity. (Courtesy of Marsha Lautenbach.)

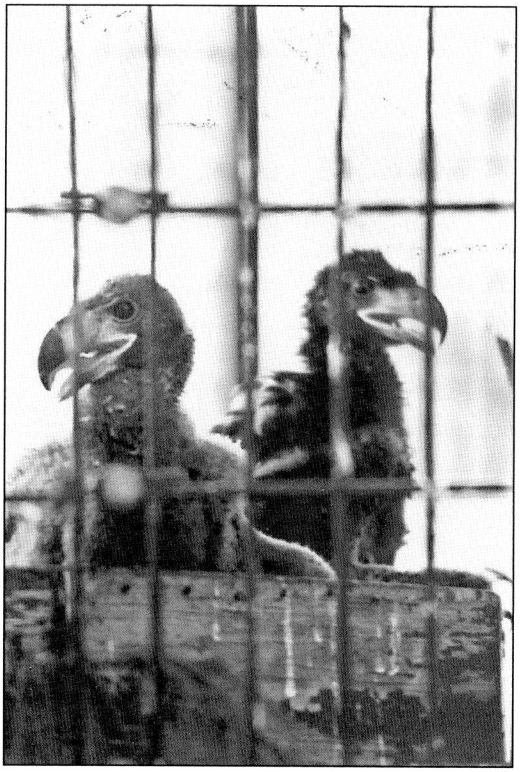

The Montgomery zoo also participated in local conservation efforts outside of the zoo. One of the most prestigious conservation efforts was their involvement with returning our endangered national symbol to the wilds of Alabama. These efforts all started with the first hatching of two bald eaglets at the zoo in April 1987. After being raised by their parents for seven weeks at the Montgomery Zoo, the eaglets were transferred on June 12, 1987, to Alabama's Nongame Wildlife Program. They were eventually released into the wild. The program had already released 10 eagles in 1987, for a total of 24 for Alabama's eagle restoration project. These, however, were the first eaglets to be entirely captive raised. (Courtesy of Marsha T. Lautenbach.)

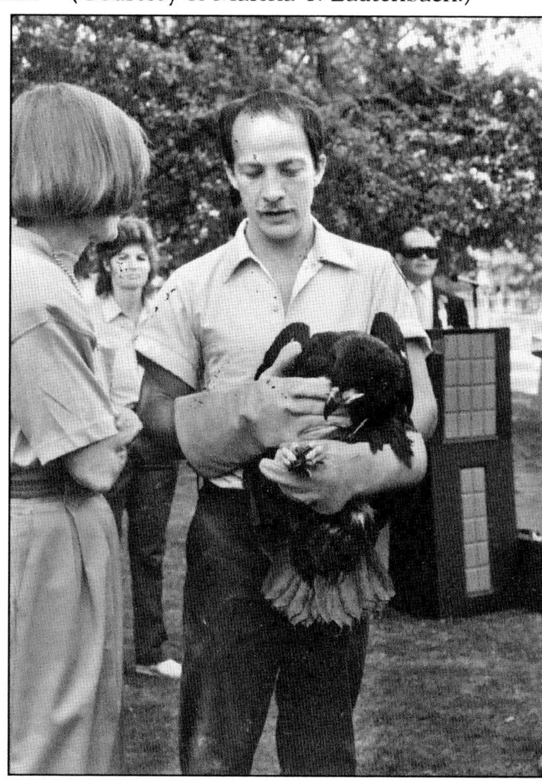

In September of 1987, the Montgomery Zoo joined the ranks of the most prestigious zoos across the country when it received accreditation from AAZPA (American Association of Zoological Parks and Aquariums), now know as the AZA (Association of Zoos and Aquariums). In the words of director Clift, this was "one of the best things that's ever happened to the zoo." The zoo had previously been inspected for accreditation in 1986 but was denied. Fixing the problems noted then, such as lack of preventative veterinary care, by contracting Dr. Douglas Carmicheal (below) to provide regular care to the now over 600 animals that call the zoo home, the zoo was able to achieve its accreditation goal. Also built was an animal care center with an operating table and emergency equipment as a precautionary measure for any unexpected problems the animals might have. (Courtesy of Montgomery Zoo.)

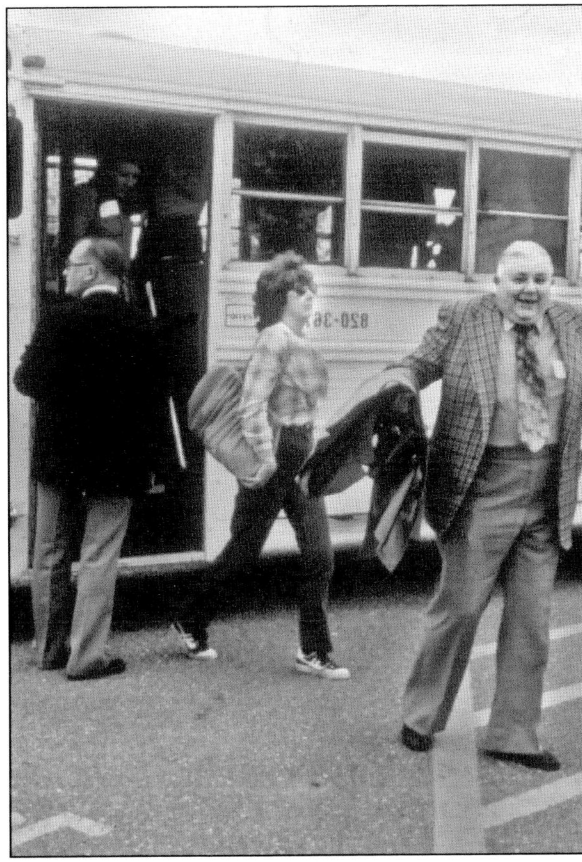

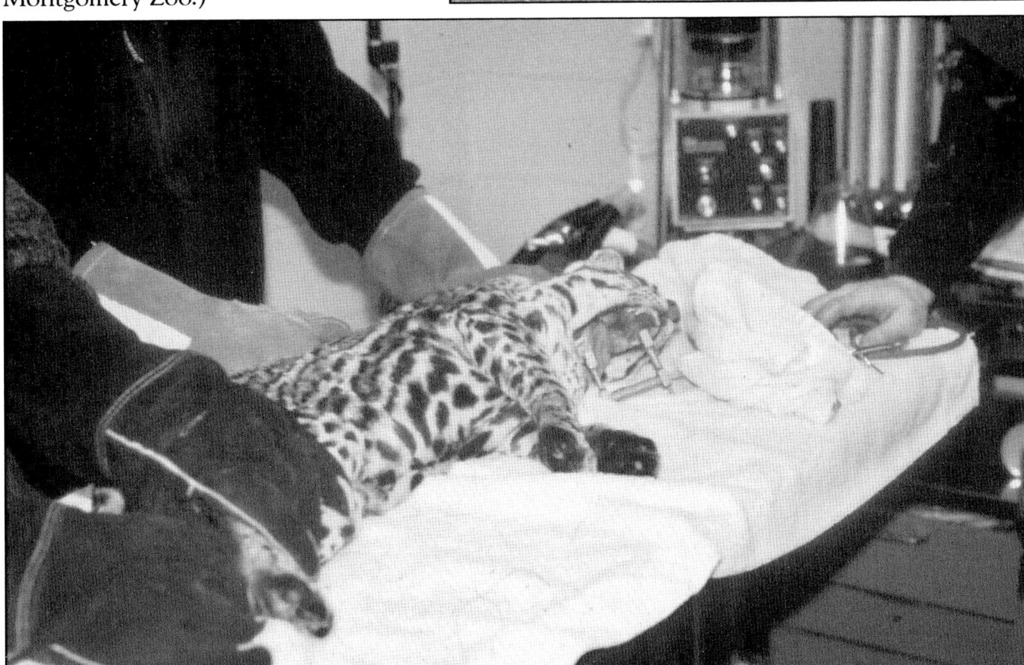

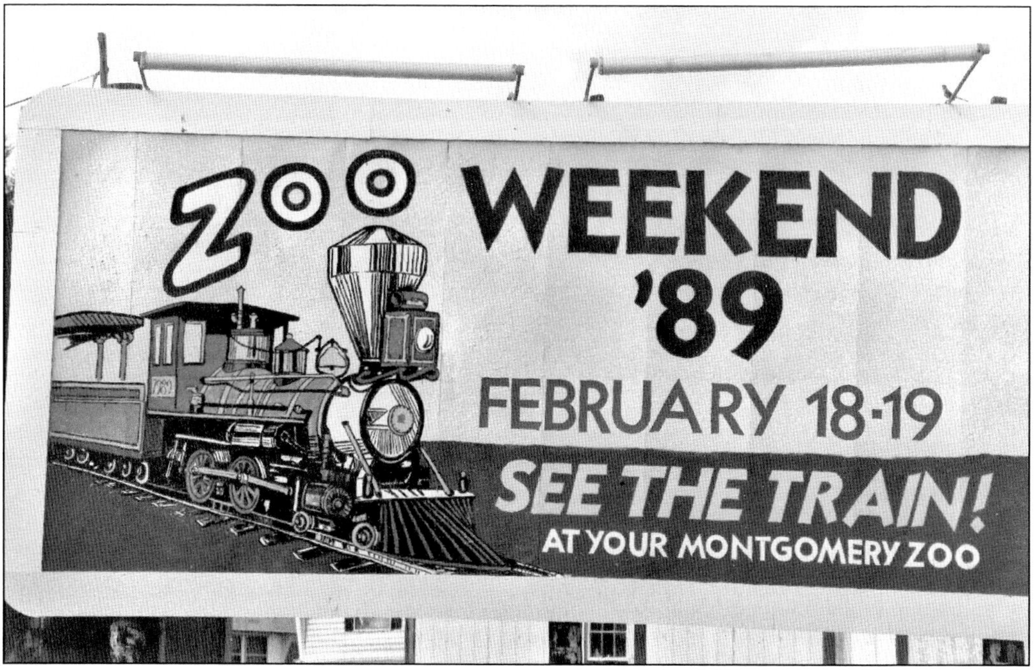

Also among the achievements of 1989 for the zoo was the Top 20 Events in the Southeast Award from the Southeast Tourism Society, which was given for the Zoo Weekend events at the zoo. Proceeds from Zoo Weekends beginning in 1987 contributed to the zoo's ability to buy a real train. A train was purchased from Pennsylvania and, although not operational, was on display at Zoo Weekend '89. Zookeeper Mike McGough was excited about the addition to the zoo. The train cost about $300,000 and was fully running by spring. Tickets were sold separately for $1 for visitors to ride the train. (Courtesy of Montgomery Zoo.)

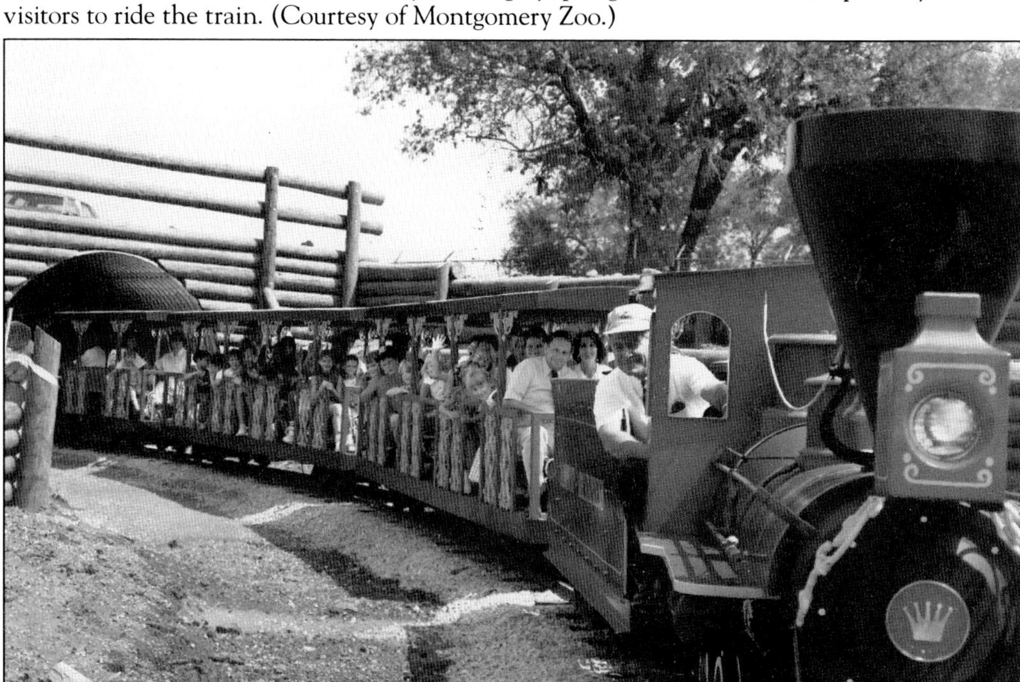

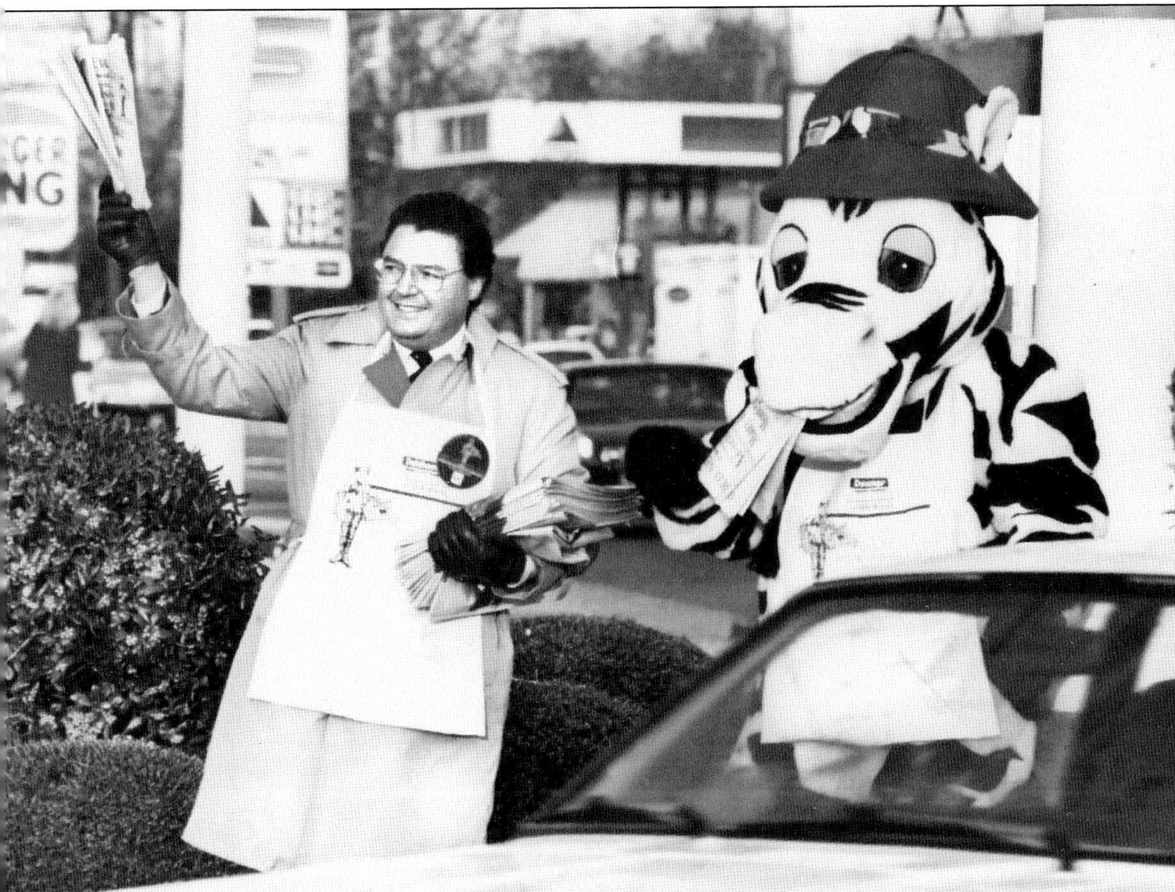

Zoo director Chuck Clift, who became blind several years ago after brain surgery, was replaced as the Montgomery Zoo director. Although no longer director, Clift remained with the zoo. William J. Fiore was named director of the Montgomery Zoo on November 6, 1990. Formerly the general curator of the Tulsa Zoological Park in Oklahoma, Fiore brought over 15 years' experience to the job. Fiore immediately began to oversee the $10 million expansion of the zoo. He had received a bachelor's of arts in zoology from Rutgers University and a master's of public administration from the University of Oklahoma. Jumping into his new position, Fiore hawked *Old Newsboy Special Edition* papers with Zoobie-Do, the Montgomery Zoo's mascot, to help raise money for the new expansion. Sponsors for the fund-raising event included the *Montgomery Advertiser* and the *Alabama Journal*. (Courtesy of Montgomery Zoo.)

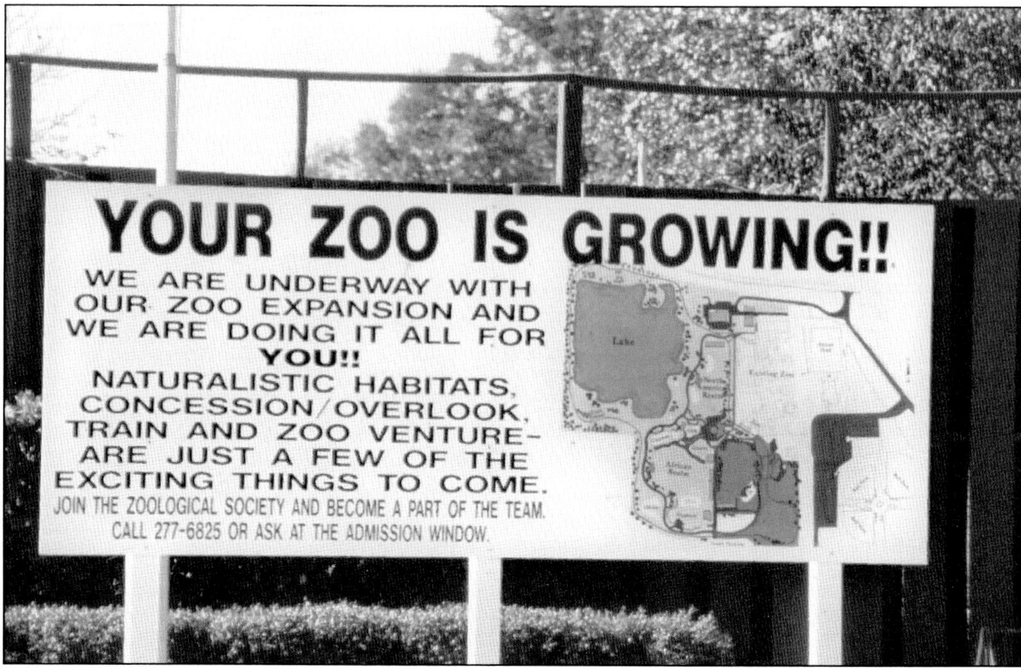

The Montgomery Zoo was in full swing, raising money to fulfill expansions outlined in its master plan. The new zoo complex, along with the Alabama Shakespeare Festival, the Montgomery Museum of Fine Arts, Old Alabama Town and the Capitol complex, was expected to establish Montgomery as a designation city for tourists from across the country. Montgomery could then reap the benefits of increased tourism, improved city image, additional city revenues, and an enhanced quality of life for all citizens. The total projected financial need for the project was $10 million. The city provided half and the Montgomery Zoo was responsible for raising the rest from private donations. (Courtesy of Montgomery Zoo.)

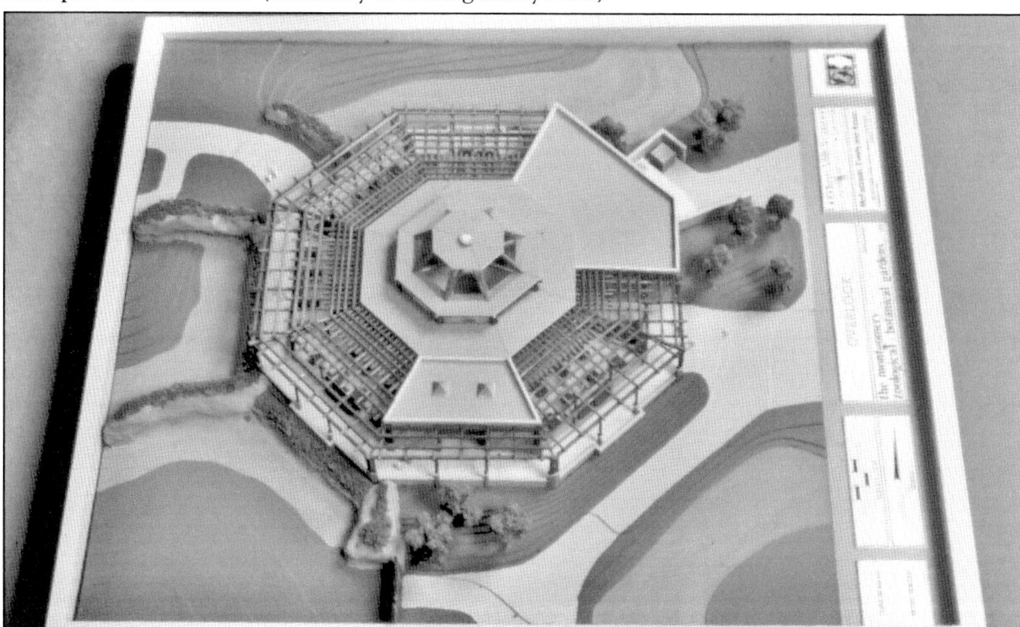

Four

Montgomery Zoo Expansion 1990–1991

In August 1990, work began on the 34-acre expansion of the zoo. The Montgomery Zoo also separated from the parks and recreation department and was elevated to city department status. Expansions to the zoo included a new activities and family picnic area, an environmental awareness center, an educational discovery area with a "show and tell" theater and a children's petting zoo, a gift and food concession area, new administrative offices, stroller and wheelchair rental facilities, and an extensive array of animals in replications of their natural habitats. Construction of the new zoo took nearly a year and was completed in late August 1991. The new zoo opened to the public in September 1991. Attendance for the zoo doubled in its first year of operation, and revenues were very strong.

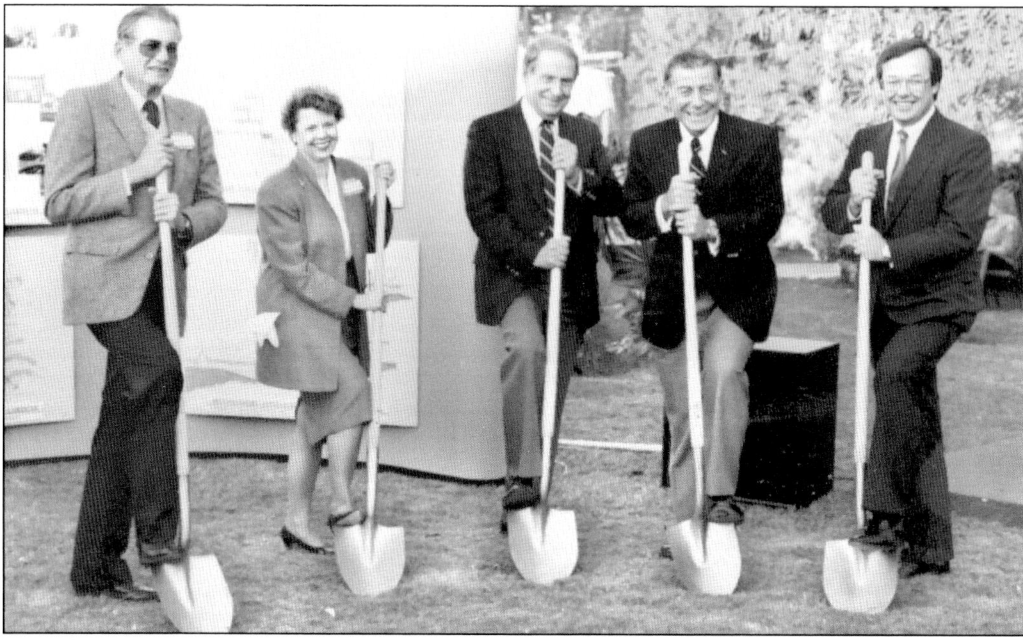

With $5 million from the city of Montgomery and the Montgomery Area Zoological Society matching funds, the expansion became a reality. The initial work included a new entrance and additional parking lots. The ground breaking was led by (from left to right) Mark Sable, president of the Montgomery Area Zoological Society; Ann Muscari, corporate vice president of development; Billy Turner, city councilman; Mayor Emory Folmar; and Randy George, president of the Montgomery Area Chamber of Commerce. (Courtesy of Montgomery Zoo.)

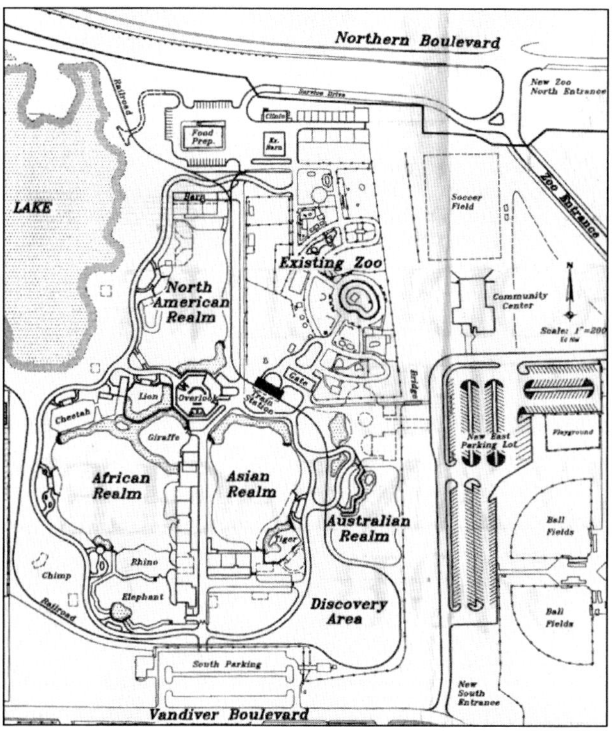

The new zoo plans consisted of five realms: North America, South America, Asia, Australia, and Africa. Phase one concentrated on the infrastructure support area (power, sewer, water, gas, and service road), train track, Australia Realm, the education discovery area, and the children's petting zoo. The goal of phase two was the Asia Realm. (Courtesy of Montgomery Zoo.)

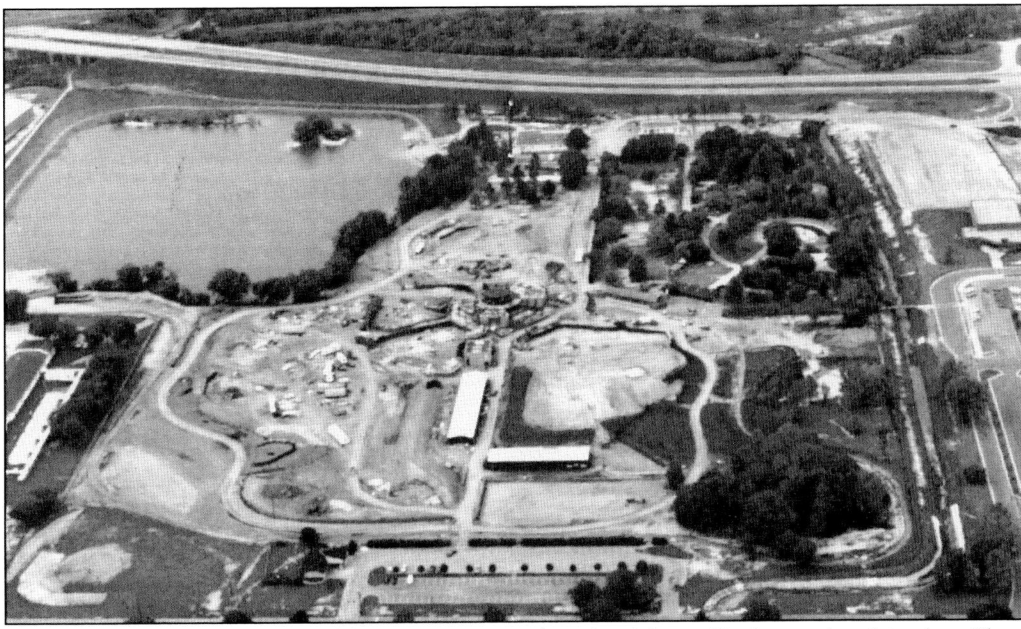

Phase three saw construction of the entrance complex, the African Realm, and parking lots. Phase four finished construction of the North American Realm and rehabilitation of the existing zoo. Completion of these additions to the zoo was another step toward making Montgomery a tourism destination city and an investment in the future. (Courtesy of Montgomery Zoo.)

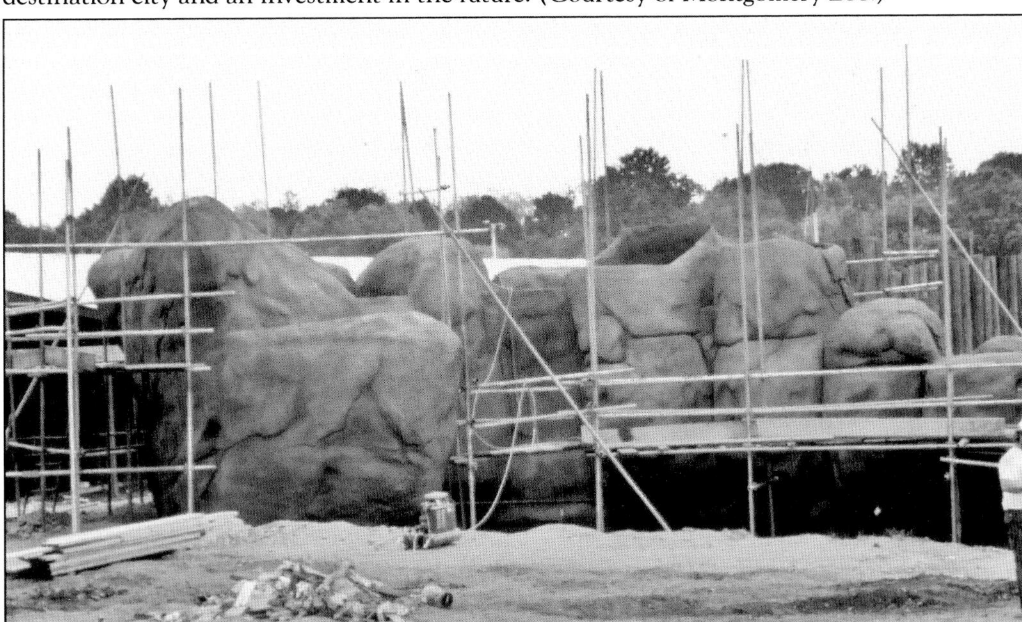

The goals of the new zoo were to provide for an enjoyable and satisfying recreational experience while revealing to the visitors their relationship within the natural environment; to create opportunities to observe and discover basic ecological principles; and to encourage each visitor to become involved in conservation and stewardship issues, programs, and practices. (Courtesy of Montgomery Zoo.)

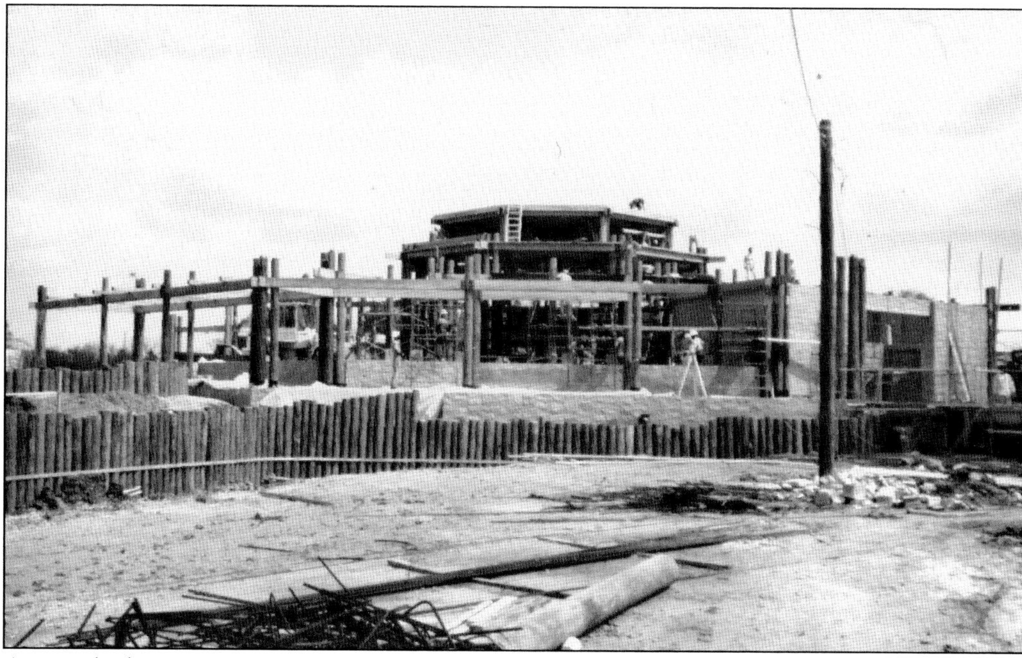

An overlook—a 14,000-square-foot pavilion that includes a full-service café—was built in the middle of the zoo, allowing visitors to have a clear view of the animals in both Africa and Asia. On September 28, 1990, community leader and president of the Montgomery Zoological Society Mark Sabel died after heart surgery. He had raised close to $2.5 million in contributions pledged for the zoo's expansion. In his honor, Mayor Emory Folmar and the Zoological Society dedicated the overlook in his memory. The Mark Sable Overlook features a plaque with the saying "he loved animals and dreamed of having a great new zoo in Montgomery. Dreams can come true." (Courtesy of Montgomery Zoo.)

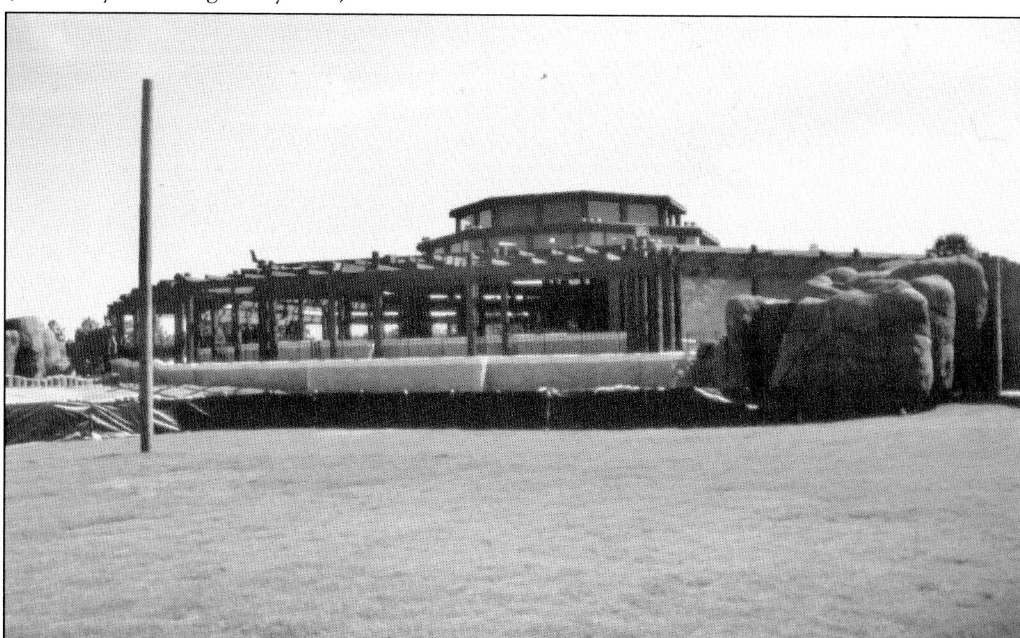

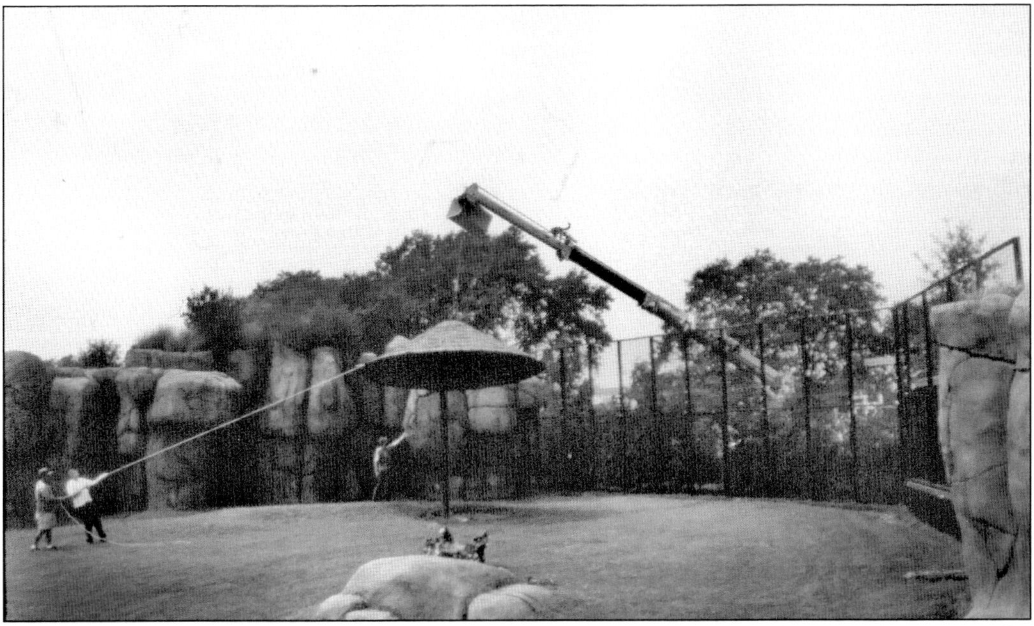

All of these enhancements raised Montgomery's environmental and educational image in the zoo world and to the visitors. The zoo was breaking free of its original image as a petting zoo and making a name for itself as a well-rounded institution that housed large carnivores as well as petting zoo animals. (Courtesy of Montgomery Zoo.)

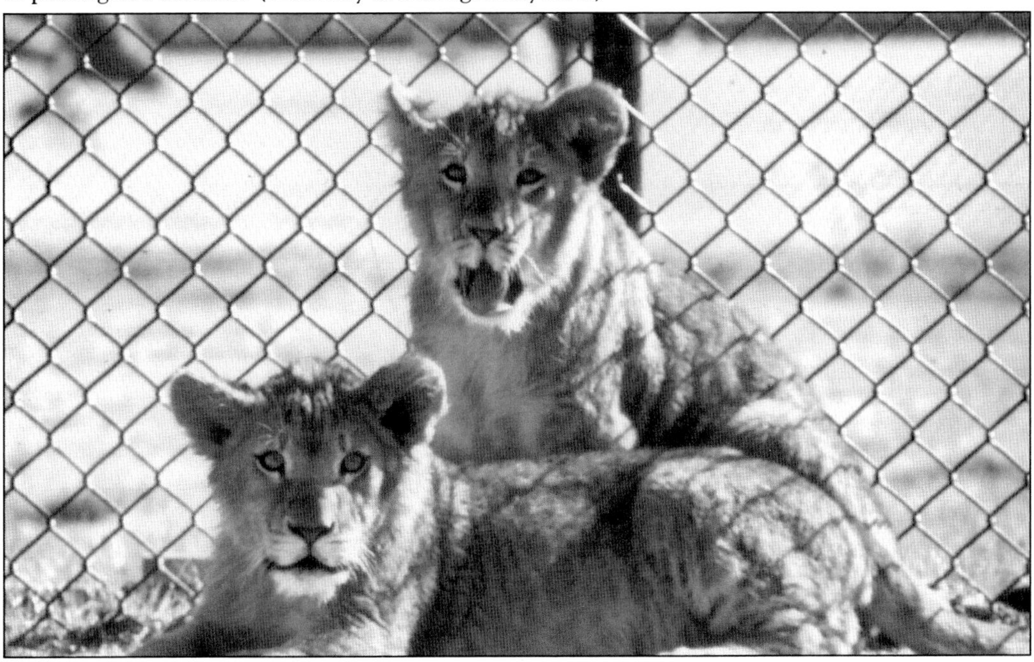

Now the public would be able to easily view the animals from the raised overlook in their naturalistic habitats. Grassy savannahs, flowing rivers, waterfalls, and rock walls all add to the viewing pleasure and welfare of the animals. The Montgomery Zoo wanted to represent the five main areas that animals are found on Earth and was successful in its presentation. (Courtesy of Montgomery Zoo.)

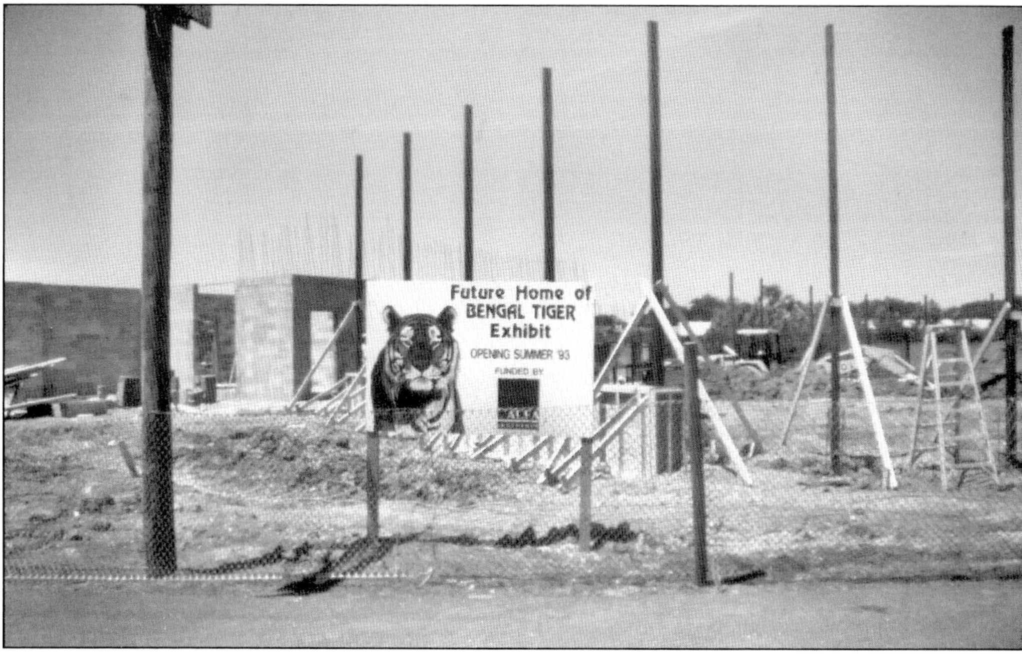

With facilities at the zoo becoming better, the programs at the zoo also received a facelift. Montgomery Zoo became a participant in the Species Survival Plan, a nationwide cooperative effort between zoos to save endangered animals. Specifically, the zoo participated in breeding and survival programs for chimpanzees and cheetahs. (Courtesy of Montgomery Zoo.)

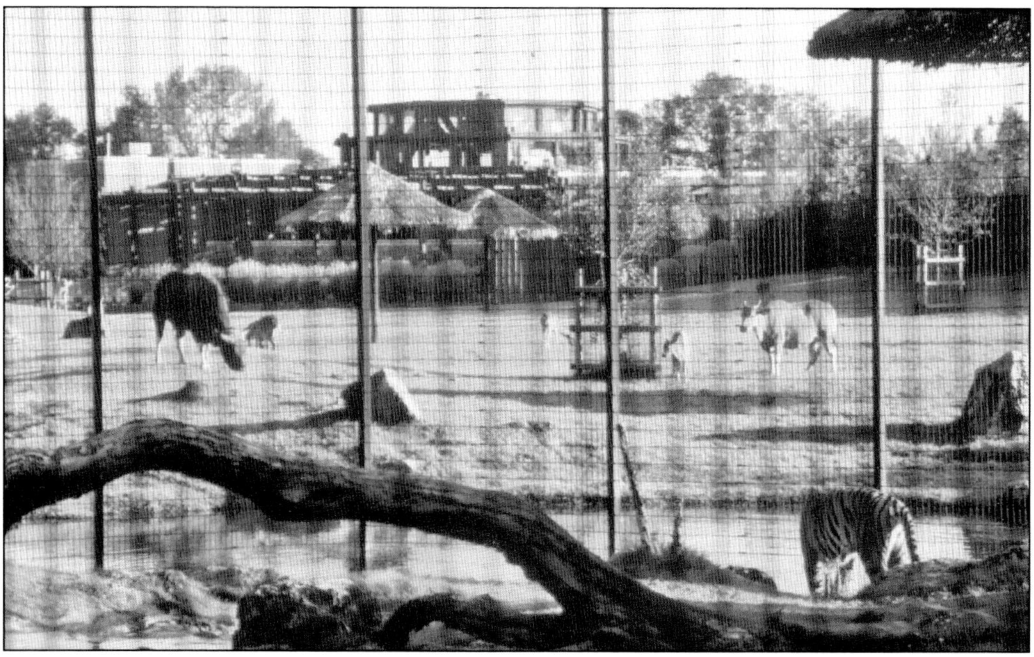

Following past zoo director Chuck Clift's original vision for the Montgomery Zoo in providing mixed-species exhibits, the new zoo embraced this philosophy in all five designed realms. Each realm contains a large range with many species representative of the area. (Courtesy of Montgomery Zoo.)

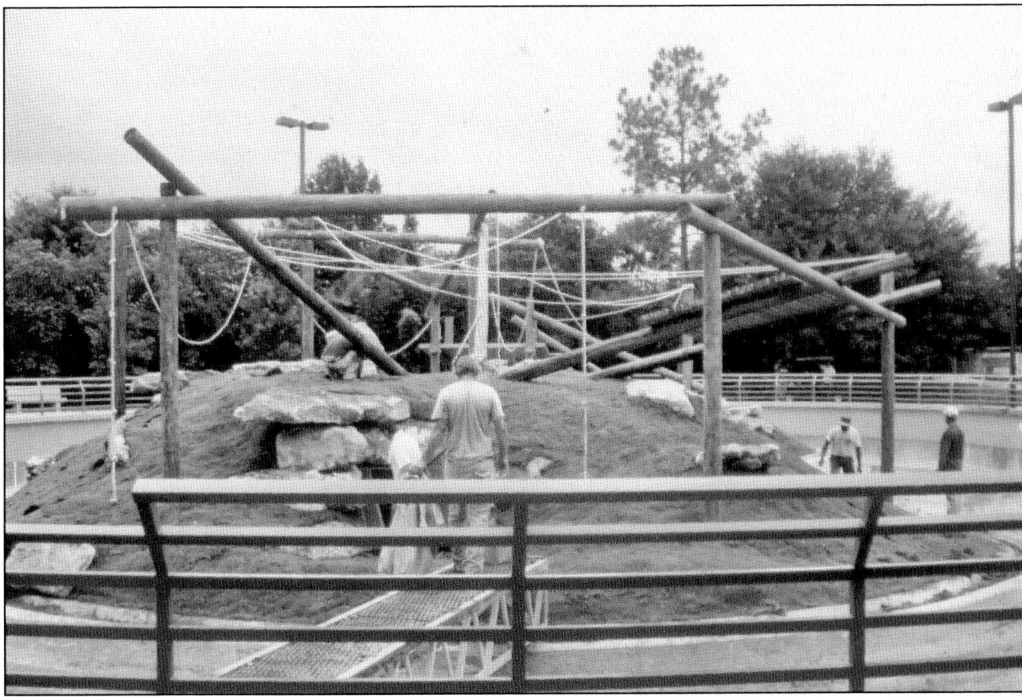

The grand opening of the zoo featured finished construction of Phase I and Phase II of the zoo's master plan. Improvements continued even after the opening with a new chimpanzee island display. Rehabilitation to the old zoo facilities and expansion of the South American Realm shortly followed. The current Monkey Island was given a facelift to become a more naturalistic setting. (Courtesy of Montgomery Zoo.)

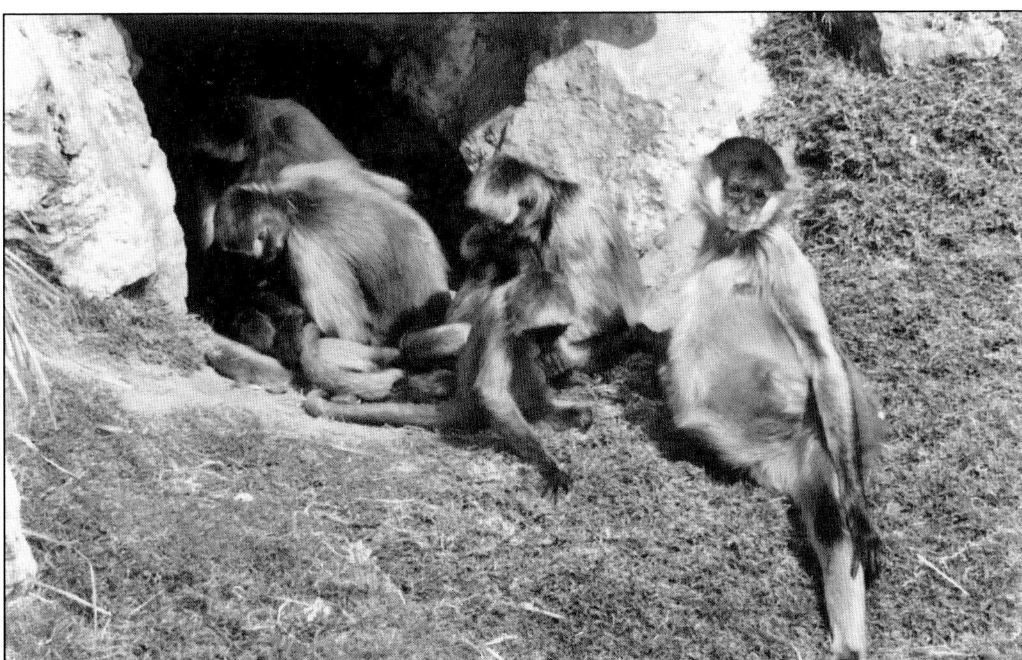

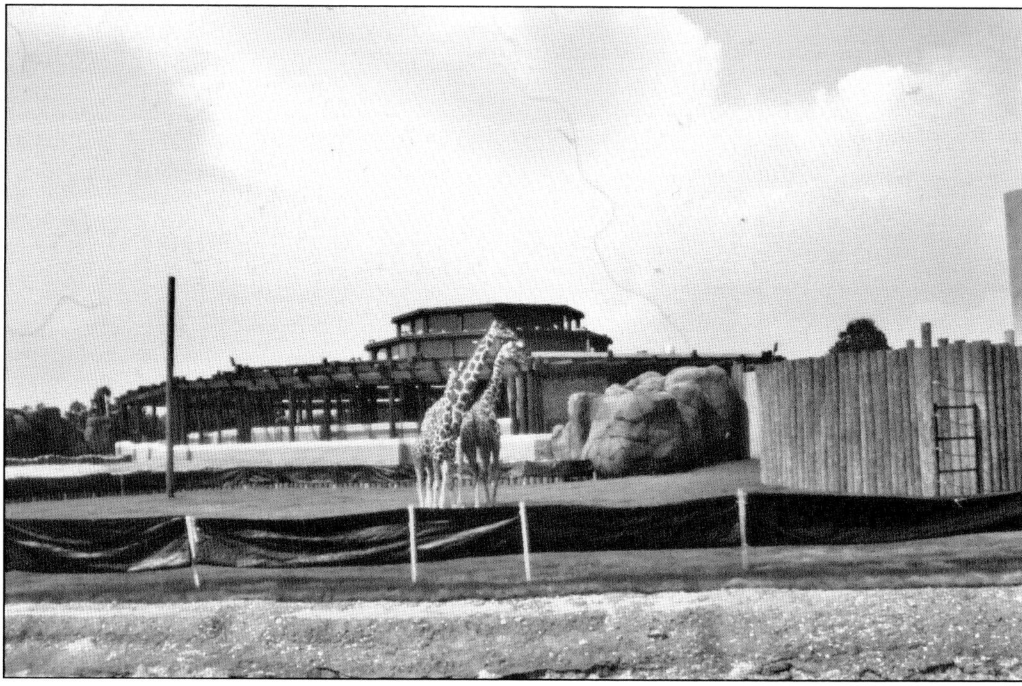

Montgomery Zoo closed on September 3, 1991, for 17 days. During the closing, the zoo relocated animals from the old 6-acre zoo to their new quarters. New animals were also brought in. Five hundred animals, representing close to 240 species, now called Montgomery Zoo home. The zoo opened on September 21 to the public. More exotic animals were on display than in the past, with more coming in the future. (Courtesy of Montgomery Zoo.)

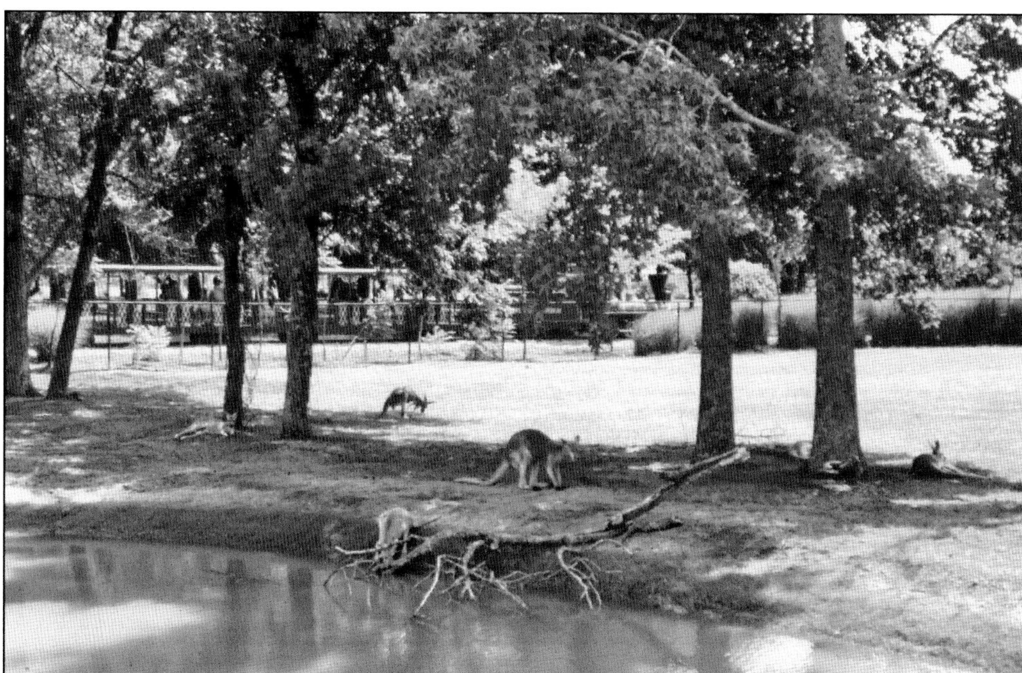

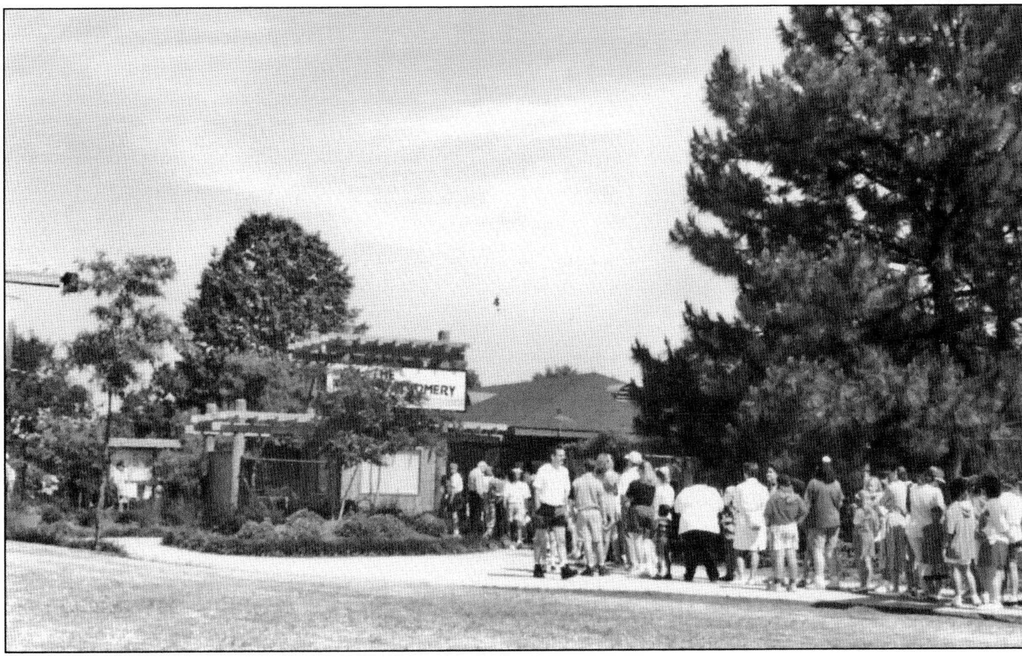

The grand reopening of the zoo was beyond successful. The new zoo was described as a "trip to the veldt and plains of African, the Australian Outback, the pre-settlement Great Plains—giraffes, lions, buffalo, all manner of other hoofed ruminants, kangaroos, ostriches, emus and goodness knows what else roam freely in different realms, separated by natural moats, backdrops and water." An invitation-only safari dinner preceded the reopening, with proceeds going toward the South American realm. Admission to the zoo was free to allow all those in Montgomery to review its new zoo. (Courtesy of Montgomery Zoo.)

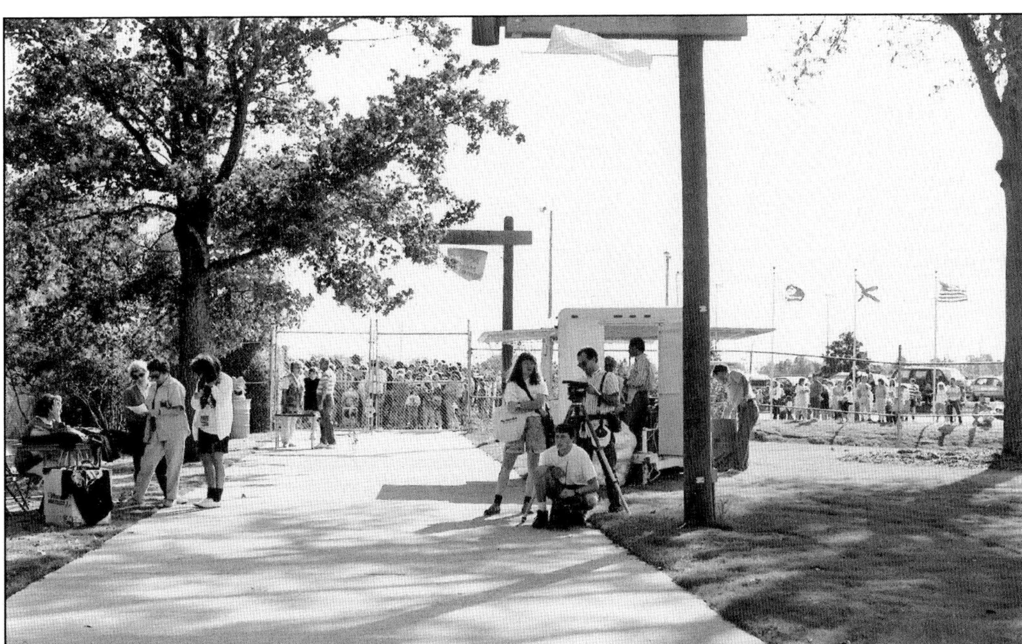

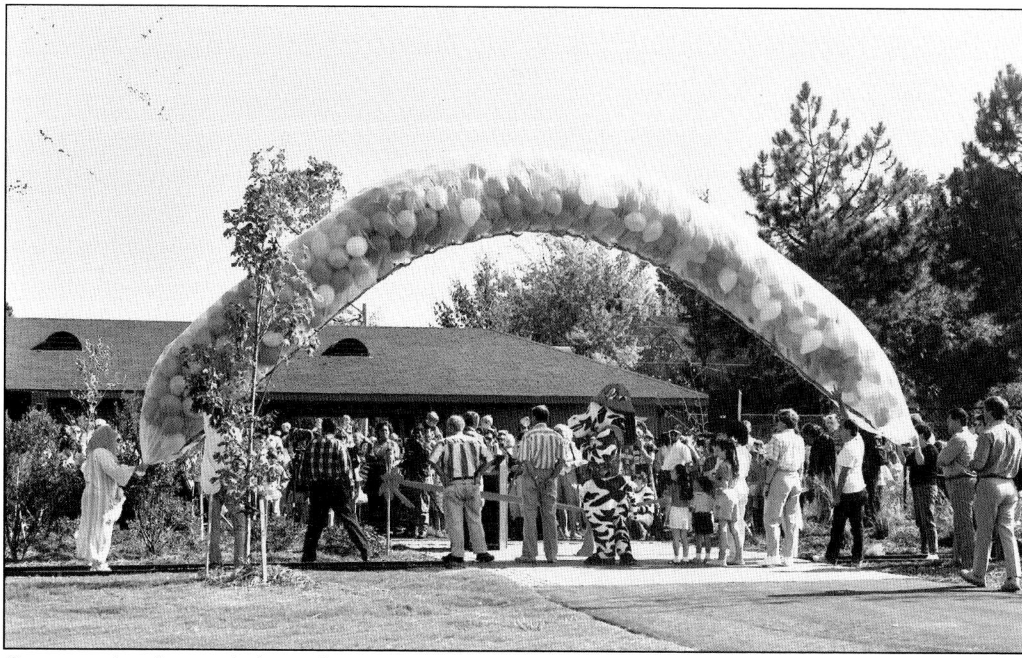

Opening day featured the Robert E. Lee High School's band and the release of 1,000 helium-filled balloons. Magicians and jugglers provided amusement throughout the day, and a storyteller entertained the children in the zoo's new overlook pavilion. Train tours of the zoo grounds were given for free. (Courtesy of Montgomery Zoo.)

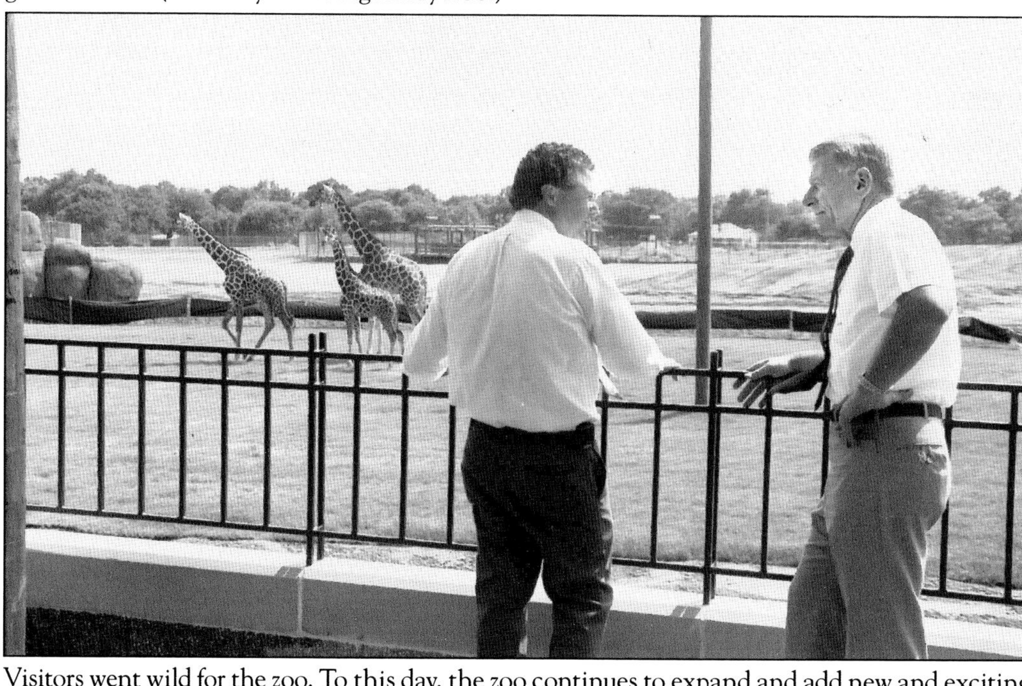

Visitors went wild for the zoo. To this day, the zoo continues to expand and add new and exciting exhibits. Not only is the zoo a haven for the animals, but it is also a learning classroom for visitors of all ages. The new zoo is a state-of-the-art facility with barrier-free exhibits featuring animals in their natural habitats. (Courtesy of Montgomery Zoo.)

BIBLIOGRAPHY

Chalk Dust. "Its Wild at the Montgomery Zoo!" Volume 1(7), April 1987.
FUTURE city guide. "The History of Montgomery's Oak Park." March/April 2007.
Montgomery Daily Advertiser. August 10, 20, 24, 1886.
Montgomery County Records of Conveyance. Book 20, pp. 519–521.
Montgomery County Records of Conveyance. Book 34, pp. 376, 377.
Montgomery City Directory 1887, p. 238.
Montgomery Advertiser. "Advertiser Files, 100 Years Ago." July 3, 1993.
Dixie Zoological Society Newsletter. "Zebra Tales." Volumes 3(5)–7(6), 1980–1987.
Montgomery Zoo Newsletter. "Jungle Drums." Volumes 1–4, 1991.